Young People, Rights and Place

Concern is growing about children's rights and the curtailment of those rights through the excesses of neoliberal governance. This book discusses children's spatial and citizenship rights, and the ways young people and their families push against diminished rights.

Armed initially with theoretical concerns about the construction of children through the political status quo and the ways youth rights are spatially segregated, the book begins with a disarmingly simple supposition: Young people have the right to make and remake their spaces and, as a consequence, themselves. This book de-centers monadic ideas of children in favor of a post-humanist perspective, which embraces the radical relationality of children as more-than-children/more-than-human. Its empirical focus begins with the struggles of Slovenian *Izbrisani* ('erased') youth from 1992 to the present day and reaches out to child rights and youth activists elsewhere in the world with examples from South America, Eastern Europe and the USA. The author argues that universal child rights have not worked and pushes for a more radical, sustainable ethics, which dares to admit that children's humanity is something more than we, as adults, can imagine.

Chapters in this groundbreaking contribution will be of interest to students, researchers and practitioners in the social sciences, humanities and public policy.

Stuart C. Aitken is June Burnett Chair and Distinguished Professor of Geography at San Diego State University. His research interests include critical social theory, qualitative methods, children, families and communities. Stuart has worked with the UN on child rights, labor and migration issues. His previous books include *The Ethnopoetics of Space: Young People's Engagement, Activism and Aesthetics* (2016), *The Fight to Stay Put* (2013), *Young People, Border Spaces and Revolutionary Imaginations* (2011), *Qualitative Geographies* (2010), *The Awkward Spaces of Fathering* (2009), *Global Childhoods* (2008), *Geographies of Young People* (2001), *Family Fantasies and Community Space* (1998), and *Place, Space, Situation and Spectacle* (1994). He has published over 200 articles in academic journals as well as in various edited book collections and encyclopedias. Stuart is past co-editor of *The Professional Geographer* and *Children's Geographies*.

Routledge Spaces of Childhood and Youth Series

Edited by Peter Kraftl and John Horton

The *Routledge Spaces of Childhood and Youth Series* provides a forum for original, interdisciplinary and cutting edge research to explore the lives of children and young people across the social sciences and humanities. Reflecting contemporary interest in spatial processes and metaphors across several disciplines, titles within the series explore a range of ways in which concepts such as space, place, spatiality, geographical scale, movement/mobilities, networks and flows may be deployed in childhood and youth scholarship. This series provides a forum for new theoretical, empirical and methodological perspectives and ground-breaking research that reflects the wealth of research currently being undertaken. Proposals that are cross-disciplinary, comparative and/or use mixed or creative methods are particularly welcomed, as are proposals that offer critical perspectives on the role of spatial theory in understanding children and young people's lives. The series is aimed at upper-level undergraduates, research students and academics, appealing to geographers as well as the broader social sciences, arts and humanities.

Children, Young People and Care
Edited by John Horton and Michelle Pyer

Youth Activism and Solidarity
The Non-Stop Picket Against Apartheid
Gavin Brown and Helen Yaffe

Children, Nature and Cities
Rethinking the Connections
Claire Freeman and Yolanda van Heezik

Young People, Rights and Place
Erasure, Neoliberal Politics and Postchild Ethics
Stuart C. Aitken

For more information about this series, please visit: www.routledge.com/Routledge-Spaces-of-Childhood-and-Youth-Series/book-series/RSCYS

Young People, Rights and Place

Erasure, Neoliberal Politics and
Postchild Ethics

Stuart C. Aitken

Routledge
Taylor & Francis Group

LONDON AND NEW YORK

First published 2018 by Routledge
2 Park Square, Milton Park, Abingdon, Oxfordshire OX14 4RN
52 Vanderbilt Avenue, New York, NY 10017

Routledge is an imprint of the Taylor & Francis Group, an informa business

First issued in paperback 2020

British Library Cataloguing-in-Publication Data
A catalogue record for this book is available from the British Library

Library of Congress Cataloging-in-Publication Data
A catalog record has been requested for this book

ISBN: 978-1-138-69772-0 (hbk)
ISBN: 978-0-367-59061-1 (pbk)

Typeset in Times New Roman
by Taylor & Francis Books

Contents

Figures

Foreword

As one of the first scholars working on the geographies of childhood and youth, Stuart Aitken has developed a global reputation for scholarship that ties together deep empirical study with formidable theoretical discussion and innovation. In doing so, Stuart's work has broached a number of issues of ongoing concern to children's lives – from children's developing relationship with their immediate surroundings to what he has termed the 'awkward' spaces of fathering. This book – which at once offers a deep critique and rich reconstitution of the notion of 'children's rights' – is no different.

Within the social sciences, the idea of (universal) children's rights has been so fundamental that it has been a cornerstone for the so-called 'New Social Studies of Childhood'. Outside academia, the United Nations Convention on the Rights of the Child has at least spawned consideration of children, their rights, and their participation in society, even if the outcomes for children (for instance) in terms of intergenerational equality or their vulnerability to climate change and other risks) have not necessarily been improved.

However, as this book so brilliantly demonstrates, the very idea of children's rights – and especially *universal* children's rights – (even if laudable) is flawed in many senses. From a scholarly standpoint, the impact of many years of theoretical challenge from various brands of poststructuralist thought, nonrepresentational theory, posthumanisms and feminist new materialisms has been to radically question the idea of the individuated human subject as a focus for social-scientific analysis. Thus – and it has to be said, particularly within human geography scholarship – the very idea of 'the child' as a bounded, identifiable rights-bearer has become virtually untenable in some circles (although not others). Outside academe, and in the kinds of geographical contexts that Stuart explores in this book, the idea of universal children's rights also appears not only contentious but, in some cases, potentially harmful. We know that in some countries, a universalizing, Westernized (and particularly Anglocentric) assumption that children should not be involved in paid labour has at best limited their ability to earn an income for themselves and their families, and in other cases has been at least a contributing factor in driving children into illegal and dangerous forms of paid work.

The difficulty, of course, is that it is for a variety of reasons – moral, political, emotional – extremely hard to entirely let go of the (perhaps) utopian vision of a universal framework for children's rights. This difficulty is compounded by the fact that – if we follow some brands of nonrepresentational or new materialist thinking – it is hard to either reconcile their theoretical propositions with a conception of 'rights', or to think what might replace a discourse of 'rights', should that be the ultimate goal. Some – not all – brands of nonrepresentational and new materialist thinking have been charged with evading these kinds of knotty difficulties. Yet this book is different. It offers a brilliant and inherently *geographical* exposition of children's rights in a range of global contexts in which the idea of children's rights has been thrown into sharp relief. For instance, the case of the erasure of a generation of *Izbrisani* youth in post-conflict Slovenia questions children's right to belong, *in place*, as well as it clearly identifies the sense in which children and young people's rights (if we are to call them such) are performative, materialized and relational: constituted through ongoing, lasting, forms of interpersonal (and intergenerational) relationships of care.

This is, then, a rich, nuanced and (in places) complex book, which sees through its ambition to bring together quite an eclectic mix of theorists in the service of holding the very notion of 'children's rights' to account. It takes significant strides towards a conception of a sustainable and life-affirming ethics of and for children, young people and their (non)human companions.

Peter Kraftl, January 2018, Birmingham, United Kingdom

Acknowledgments

As in all my writing, I am indebted to numerous people and most of all to my family, friends and close colleagues. Specific acknowledgement goes to Peter Kraftl and Giorgio Hadi Curti who read an earlier version of the completed manuscript. I am particularly indebted to Peter who returned to me in a surprisingly short amount of time a close reading of the work. He was particularly astute in opening up some of the postchild literature in early childhood education about which I was unaware. Giorgio was delightfully honest in questioning how I could get Spinoza so wrong. I also want to thank Jasmine Arpagian and Adriana Cordeiro for working with me on the Romanian and Brazilian cased studies respectively, and to generously contributing portions of their own work to this book through our collaborations on other projects. Also, in addition to the footnote in Chapter 4 I must profusely thank Ines Hvala for her contributions to the Slovenian fieldwork, which not only included translation and transcription but also sleuthing out participants for the project and driving with me all over the country. Ines was industrious, stalwart and a wonderful companion. Finally, a portion of the DREAMers discussion comes from as yet unpublished work by Ala Sirriyeh, whom I met for the first time at a conference in September 2017 and who kindly allowed me its use.

1 The postchild and the law thing

The Faroese believe that childhood on their islands is different from most other places. This small, isolated archipelago in the North Atlantic has a population of just over 50,000, half of whom live in the capital Törshavn. The islands were inhabited relatively late on as these things go, attractive in the first instance to 6th century Irish monks looking for solitude. The islands got their first television station in 1985, again much later than most other places. More recently, things began to speed up; the Faroe Islanders created one of the best infrastructures in Europe with an enviable road, bridge, tunnel and Wi-Fi system linking all the larger and more populated places that were heretofore only accessible by boat. Education levels are exceptionally high, and the University of the Faroes boasts renowned graduate and under-graduate programs. Researchers at the university, interviewing and surveying Faroese adults and children alike, note an unprecedented amount of freedom for children (Gaïni 2013; Hayfield 2017). Young people, between 16 and 21 years of age, report much higher levels of feeling good or very good about their mental health (~75%) when compared with other Nordic countries, and 88% were content with their lives (Guðmundsdóttir et al. 2010). Youth unemployment is less than five percent (Petersen 2016). More children are born in the Faroe Islands today than in the last one hundred years, and Faroese women continue to have more children on average than in most Western European countries (Johannesen 2017).

In no way do I want to romanticize Faroese childhood, as it is, like every-where else, replete with classed, raced and gendered issues, but I do want to offer it as a Foucauldian 'affirmative critique' of an affective context that asks questions about new forms of society and governance (cf. Deleuze and Guattari 1994; Foucault and Rabinow 1997; Staunaes 2016). What is it about the Faroe Islands that is so amenable to children and young people? Why do young adults who leave the islands invariably return to have their own children or to retire (Hayfield 2017)? What is so compelling about the natural and social environment, and the cultural and political milieu, that caregivers want to raise their children there, and children want to stay? Here is the halting story that I tell myself as I try to answer these questions during a four-week visit in Spring 2017.

Figure 1.1 Typical village nestled between two hills on a northern Faroese island
Source: author

The Faroese natural environment – oceans, cliffs, wind, rain, cold – is omnipresent and brutal. The landscape is spectacular (Fjord coastlines, thousand meter cliffs, waterfalls, and rugged mountains), but it is also fore-boding (the ever-changeable weather and stormy seas); the islanders' national imagination is closely linked to nature and surviving against its daunting presence (Figure 1.1). Now, the last thing I want to do is to push some needed link between children and nature as a panacea for society's ills, which seems like a popular thing to do and with punishingly essentialist outcomes (cf. Louv 2005), but Faroese children nonetheless connect with nature at an early age. They learn about the dangers of oceans and cliffs, and rarely wander off and get lost or put themselves in danger. Nevertheless, wander they do, far-and-wide, and for the most part without adult supervision. Although beset by storms, the Faroes have a relatively mild winter and cold summers and so nature does not necessarily curtail mobility. Children as young as six and seven years of age come together for picnics on beaches and in the hills unsupervised by parents or caregivers (Figure 1.2). They walk to-and-from school often over long distances in most kinds of weather. Groups of young people roam around the capital Törshavn, hanging out, having fun and generally not getting into too much trouble. It is as if young people have a right to free association with, and movement in, the spaces through which and from which they grow. With that said, along with Peter Kraftl and his colleagues, I worry that although notions like free association, mobility and movement are important and ideas like 'nomadism' are potentially liberating they may well be enervating in a boarder sense when thought of along with classism, ageism, sexism, racism, and in particular geographic contexts (Kraftl 2006, 2013a, 2013b; Horton and Kraftl 2006; Barker et al. 2013).

Figure 1.2 Young Faroese children play unsupervised on a beach
Source: author

Clearly, not all children in all places at all times have these freedoms, and not all those that do, find them beneficial. One of my intents in the pages that follow, then, is to think more deeply about what is meant by freedom, how freedom relates to rights, and the ways that what we think of freedoms and rights might map onto children.

The Faroese community is close-knit and family-oriented. The Faroese children on the beach and in the hills, and the young people hanging out in Törshavn, are free, to a large degree, and also they are known; it seems that everybody looks out for everybody else. There is a certain comfort for many people who are born and raised in the Faroe Islands that relates directly to the close social and familial ties, and there, finally, we find something of a rub for young people. Until recently, the one major complaint of young Faroe Islanders was the lack of privacy. It was difficult to get away, to have private moments, and everybody seemed to know your business. Then came the Internet and social media, and a very good Wi-Fi structure. Armed with laptops and smart phones, young Faroese can now choose to connect with each other in ways that adults are unable to anticipate, oversee, or coerce. If they choose to, young people can create private moments and spaces away from peers. This technological appendage re-defines the context of Faroese childhoods in mostly good ways, at least according to the young people with whom I talked. The reframing of identity through technology is recreating Faroese childhood in important ways. Young people connect to each other, to family-members, nature, villages and a national idyll, to smart phones, technology, social media and a larger global presence in ways that suggest something more than the childhood their parents knew.

Yet, there is something more going on here, something beyond the individual child's use of social media and the Internet. Faroese society spends a lot of time and resources supporting its children and their endeavors. It ratified, embraced, codified and embellished the proclamations from the United Nations Convention on the Rights of the Child (United Nations 1989). Like many Northern European nations, they have an ombudsman whose job is to engage young people, listen to their issues, and move forward with them on what is important. Save-the-Children is a prominent institution on the islands. Inculcated to the Faroese people through their laws, but also dispositions, are the rights of children for provision, protection and participation. Young people get more than a say in the matters that concern them. At one level, this is simply another Scandinavian country that takes pride in giving children the space to grow and perhaps become something different. The focus of its laws on child rights and freedoms mirrors what can be found in Finland, Norway or Sweden. Nevertheless, the story I tell myself is that there is something more here, on these remote North Atlantic Islands with their relatively young society; something perhaps to do with pride in an ancient democratic parliament.[1]

At the risk of romanticizing the evolution of one of the oldest places of government in the world, I think that there is something quite wonderful about the way that the Faroese *Løgting* – literally 'Law Thing' – pushes towards bringing citizens (young and old) together with the intent of devolving "power and assets to the people ... to build a society characterized by resilience, solidarity and self-sufficiency" (Johannesen et al. 2017: 1). This is certainly true for most of the Faroes' recent history, and I like to think that it was a prevailing wisdom over the *Løgting's* long existence (beginning perhaps as early as 650AD).

a   b

Figure 1.3 (a) Groups of unsupervised children play in Törshavn, beside (b) the Tin-
ganes, or ancient government buildings (still in use today) that host the
Faroese *Løgting*
Source: author

The Faroese focus on a society that is resilient, concurrent, and self-sufficient suggests an ethic of sustainability. I spend some time with this concept in the pages that follow, and it is the central focus of my conclusion, but at this time it is worth a cursory definition. By an ethic I mean, in the Spinozan sense of the term, (i) an infinite set of connections that relates everything to everything else, and (ii) the outcomes of those relational encounters (Deleuze 1990). When bodies or ideas encounter other bodies or ideas there is an outcome, which may result in a powerful combination or it may result in an erasure of one or more of the bodies or ideas. Sustainability, as I use the term here, is not the popular idea of maintaining what we have into an unknown future, but rather the idea of gaining potential in the perpetual moment of the present. This form of sustainability enables a fuller potential right now through increased capacity. In the same way that a capacitor in an electrical circuit stores energy, so too this form of sustainability builds potential through increased capacity. Sustainable ethics, then, are not about establishing and maintaining rights but rather they are about fomenting and pushing change. Under the aegis of these ideas comes a focus on children and youth. Sustainable ethics build the capacity of young people to change themselves and the world in which they live. One of the outcomes of this focus is to dispel the seeming right of young people to free association with, and movement in, the spaces through which they grow and develop, as noted above, for something more important: giving them the capacity to recreate those spaces and thereby recreate themselves (and us).

The Faroese story I am telling here has a romantic, rosy hue. I was on the islands, talking with academics and interacting with young people, for only a short period. If my story smacks of utopianism and impossibility then I am guilty as charged, but I hold with David Harvey (2000: 255) that as "active subjects, consciously pushing human possibilities to their limits," we can create an aspirational vision of what the future holds so that we may live better today. As Harvey (2008: 23) notes in later work, rights are about gaining our "heart's desire." What the Faroe Islands suggest to me is the hope embedded in an aspirational vision of sustainable ethics emanating from thinking about children's rights in a different way.

It is my purpose with this book to think about sustainable ethics as a corollary to, and amplification of, universal child rights. Universal rights are necessarily global and abstracted from local contexts. Preconceived notions of a generalized human condition particularize and delimit universal child rights. This intransigence of universal rights tends to stultify discussion, whereas the nomadic fluidity of sustainable ethics keeps conversations going, and is less likely to foreclose upon children's everyday politics. Sustainable ethics are not necessarily about local rights (although they can be), but they are about something that is locatable and everyday. My purpose in what follows is the try to elaborate the sustainable ethics that I sense are part of Faroese child cultures and politics. This is a difficult task because the sustainable ethics that emerge from the chapters that follow are not necessarily idiosyncratic or peculiar to particular places, but this is from where they seem to arise. By the end of this book – once I've weaved ethics, laws, rights, and sustainability with the precarity and unpredictability of children and young people – my hope is that we glimpse a postchild moment, which is a moment that moves us through and away from the invention of childhood and what came after.

The invention of childhood and its repercussions

If it did nothing else, Philippe Ariès' *Centuries of Childhood* (1962) established that childhood was a contestable category and not always as we have come to know it in contemporary Western society. There is general agreement with his argument that industrial capitalism promoted childhood as a separate sphere of experience, and from that time, the idea of children's spaces and development started to count in important ways that grew in the decades that followed. He famously points out that prior to industrialization children "did not count" (1962: 128) because prevailing high levels of mortality did not inspire connection to a being whose hold on life was quite tenuous. Ariès (1962: 128) prevaricates that "the idea of childhood is not to be confused with affection for children: it corresponds to an awareness of the particular nature of childhood, that particular nature which distinguishes the child from the adult, even the young adult." Although criticized for flawed methods (Stone 1974), inappropriate assumptions (Wilson 1980), and a lack of consideration about child dependencies and relations (Pollock 1983), Ariès' work nonetheless set the stage for several decades of work on understanding the ways children and youth are socially constructed and different from adults.

David Oswell (2013: 9–10) argues that sociologists and historians use Ariès work to promote three reductions that have been highly significant in delimiting questions about children's agency and rights: (i) that childhood is an historical invention, (ii) that childhood is socially constructed, and (iii) that childhood constitutes social and spatial divisions between children and adults. He notes that these assumptions highlight categorical forms of conceptualization and is skeptical as to whether they serve a broader understanding of children, their agencies, and their rights. I am going to argue further, in what follows, that perhaps they do such a disservice to the relations between children, adults and society that maybe it is better to re-think how we use children's agency and rights. The assumptions that come from Ariès' work (and certainly from geographers who use his work) also highlight a somewhat problematic and particular construction of space along with divisions between children and adults, and within communities.

Two questions of some importance (and a third to follow in a moment) arise from Oswell's critique. What is the "particular nature" of childhood that distinguishes children from adults (Ariès 1962: 128), and how immutable is it? To the degree that these questions are unanswerable as posed, it is appropriate to raise some of the ways reliance on Ariès' work reduces how we come to understand children as we try to grapple with the possibility of a postchild moment.[2]Ariès (1962: 411) argued that pre-industrial "collective life carried along in a single torrent all ages and classes," and it is not inappropriate, as we begin to consider what a postchild moment looks like, to question why we stopped thinking in this way. Pre-industrial community life was public life, and the function of families was primarily to ensure the transmission of property and names. The vantage of the individual, family or community was inseparable in pre-industrial experience, and from the time they were weaned, children were part of that inseparability. With industrial capitalism, a public and private sphere was seen to emerge, and children were very much part of the protected, private sphere. With this notion of divisions, Postman (1982) argued compellingly that the decline of American (and European) childhood as a protected family space began in the 1950s with increased institutional control of children's lives (especially through media but also with changes in the educational system). From this time, childhood extends visually, spatially and temporally and by so doing it produces a situation in which there is no apparent or clear-cut division between childhood, youth and adulthood. In the latter half of the 20th century, the inculcation of family values in the home, and community values in the school and church, gave way to the uncontrolled invasion of children's minds and lives by neoliberal educational practices, market-driven media images, and globally circulated mandates for development. This foray, argue Postman (1982) and his followers (cf. Cunningham 1995) signals a loss of innocence and the disappearance of childhood as a cherished institution: "We are left with children who rely not on authoritative adults but on news from nowhere. We are left with children who are given answers to questions they never asked. We are left, in short, without children" (Postman 1982: 90).

Whether this transformation is real or not, I think that it is worth looking at the moral implications of an imagined loss of the grounding and authority as to what constitutes childhood; what is the fallout from the seeming disillusion of identifiable distinctions between childhood and adulthood? It seems clear that academic and policy concern from the latter half of the 20th century onwards sustains an emphasis on children's rights and agency (Adams et al. 1971; Archard 1993, 2004; Oswell 2013). These rights conjoin with a weakening of parental and caregiver authority (Elshtain 1990; Wood and Beck 1994), and the growth of a culture that emphasizes the personality and agency of the individual child and parent (Frønes 1994). It is no surprise, then, that there is a strong connection with how we think about children, and how we legislate on their behalf. What is of more concern for what I want to write about in this book is how these representations and actions tie to rights, individualism, and global neoliberal capitalism, while missing other more subtle connections to the material and non-material aspects of young people's day-to-day lives.

A third question worth posing at this time, then, is what exactly do we gain from a relational study of children that focuses on the material and non-material associations from a postchild perspective? Ariès and Postman were intent upon uncovering the relations between children and adults, which led to a consideration of the ways children are socially constructed, and work since then has attempted to understand the complexities of intergenerational relations across more than one generation and more than one place (Katz and Monk 1993; Kjørholt 2003; Abebe and Kjørholt 2013). Peter Hopkins and Rachel Pain (2007) expand this further to encompass a relational geography of age that simultaneously considers intergenerationality, intersectionality, and lifecourse. The postchild moment moves this trajectory further out and further in by considering a multiplicity of relations across time and space, including the material, the living, the cohabiting, the distant, the spiritual and the ephemeral. From this, what I gain from thinking about material and non-material relations is a whole series of inspiring new questions about children and young people.

The questions that I find interesting do not relate to the nature of childhood or its disappearance. Nor do they relate specifically to social constructions of childhood, and how those constructions translate into rights-based agendas. What happens if we return to thinking about children relationally in a coherent and unromantic way? What happens if we do away with any kind of social construction of childhood? How can we rethink the social and spatial divisions between children and adults so that we no longer place young people in rarified 'child only' spaces? What if we try to rethink children's (and our) spaces of existence as fluid, relational, mutable and undividable? What happens to child rights if we try to move through and beyond universality and individuation? What happens if we let go of the nature of children, their subjectivities and political identities? What happens when we think of children as more-than-children? Is it possible to think about children's humanity as something different from what we conceive for ourselves as adults? What opens up if we dare to admit that children's humanity is something more than we, as adults, can imagine?

Postchildhood

> The status of children is a chapter apart, from forced labor, to the child-soldier phenomenon; childhood has been violently inserted in infernal cycles of exploitation.
>
> (Braidotti 2013: 112)

In her acclaimed book, *The Posthuman*, Rosi Braidotti (2013) posits only this one sentence about children. She embraces a feminist politics of location and relationality to move our thinking beyond postmodernism and post-structuralism, beyond ideas of political identity and subjectivity, to something that is posthuman. Where Braidotti leaves us is within grasp of a chaotic and vibrant cosmic energy, which she describes through what she calls vital materialist methodologies. It is an inspiring thesis; Braidotti recognizes that all things human, non-human and more-than-human connect in wonderfully complex ways, without reducing or deconstructing their complexities, or framing them as subjectivities, identities, or social constructions. She does so while outlining a way forward for the sustainability of our planet inspired by post-colonial and race studies, as well as feminism and environmentalism. It is important to recognize that post-colonialism and feminism provide coherent and locatable political bases for Braidotti's posthumanism, which sets it apart from other work that is criticized for obfuscating politics (cf. Barad 2007; Murris 2016). Outlining contemporary debates on the posthuman, Braidotti begins by laying to rest the legacy of Enlightenment rationality, reason and dualistic thinking along with linguistic, representational, psychoanalytic and post-structural ways of trying to come to terms with the complexities of our lived and non-lived world, our humanness and un-humanness (and inhumaneness).

Throughout this work, Braidotti mentions the status of children only in the one sentence that I use as epigraph for this section, with the admonition that children require a fuller treatment; a treatment that raises the specters of violence and exploitation. My project in the pages that follow is to provide some of that treatment, in a partial and halting way. Throughout the book, I highlight violence towards, and exploitation of, children and young people, and comment on how a focus on universal child rights attempts to counter this in mostly problematic ways. With a beginning focus on child rights, I make a push in the balance of the book towards something more radical and less tied to universalism and individualism, landing softly on what Braidotti calls a sustainable ethics that very much relates to what I describe earlier, which for her (and me) emanates from a locatable feminist politics. It is from this location, this place of becoming, that young people create and recreate their spaces, and themselves. It is a place that realizes child actions and doings that may not resonate with adult sensibilities; it is about events and moments that enable surprise, dislocation, and radical transformation. It is a place through which young people become other, and different from us. It is a place that enables the world to move on in perhaps small, but nonetheless positive ways.

As pessimistic as the stories that unfold in what follows may appear, I argue that there is hope for children and young people from locatable feminist politics and sustainable ethics as articulated through posthuman debates. Of particular interest is a rekindling of the idea of a codependence between children and adults but not in the pre-modernist sense of a *Gemeinschaft* community, or the modernist context of young people as the future of the world, both of which merely recognize a particular and limited idea. Rather, I want to drive cooperative dependence as an elaboration of, and parallel process to, the moral philosophy of universal rights. This is a push towards, and a move with, sustainable ethics. For Braidotti (2013: 93–4), an ethics of sustainability is found only in a grounded, situated, and very specific feminist politics, and those ethics derive from what she calls *zoë*-centered egalitarianism. *Zoë* is similar to what Giorgio Agamben (1995) describes as bare-life, stripped of the pretensions of hierarchical subjectivities, binary identities, and political representations. For Agamben, bare-life dwells in a state of exception in a modern political system that is intent upon depriving certain groups of basic rights. It is the denial of citizenship and rights found in exceptional places such as Nazi concentration camps, Abu Ghraib, Guantanamo Bay, contemporary European and Australian refugee camps, and detention centers for unaccompanied child migrants in the USA (Aitken et al. 2014). Aihwa Ong (2006) extends this idea of a state of exception beyond the confines of gated camps to neoliberal contexts where subjects lose rights in a more insidious way and devolve into a state of precarity. There is nonetheless a kernel of hope in the depths of exception. Noting that bare life resides in the fuzzy and inculcate spaces of modern life alongside state authority, Agamben (1995: 9) argues that in its separateness – "a hidden foundation on which the entire political system rest[s]" – bare life frees itself and "becomes both subject and object of the conflicts of the political order, the place for both the organization of State power and the emancipation from it."

For Braidotti, pushing Agamben's ideas a little further, *zoē* is an antidote to the measure of a human provided by models such as Leonardo da Vinci's Vitruvian man, the humanistic idyll, or majority/minority subjectivities. Braidotti's posthuman thinking is trying to dispel these ideals and binaries, and to the degree that *zoē* is what remains when all things 'human' are removed from a person, it provides a useful basis for understanding the posthuman. For Braidotti, *zoē* is not about some basic animality, it is about the relations between children, adults, women, men, technology, institutions, animals, microbes, stones, and the mortar that make us who we are, and more-than-human.

What does this hyper-relationality mean for children and young people? The rise of postchild sensibilities – in conjunction with posthumanist thinking in general – recognizes young people's relations, ambiguities, dependencies, autonomies, and politics. It understands that at any one time and at any specific place the actions, practices and politics of young people are an assemblage of relations with other young people, technologies, adults, animals, and materialities that cast doubt on the nature of being and becoming. Postchild sensibilities recognize further that efforts to represent young people, or to frame their experience and act on their behalf, may show up at best as an extension of adultist or colonial thought, and at worst as a form of repression and violence. There is gathering interest in what might be called postchild methodologies focusing on, for example, language and the material world (Rautio and Winston 2015), sexuality and the classroom (Blaise 2005), play and picturebooks (Murris 2016), and indigenous politics (Taylor et al. 2012), but to date there has been little consideration of how a postchild perspective might help us with children's rights and freedoms.

In what follows, then, I use posthuman thinking as a challenge to the efficacy of children's rights and while not dismissing their importance to theory, practice and policy, I ask if perhaps there are better practices and politics through which to improve young people's lives. It seems to me that there is possibility for a different moral philosophy of rights, which recognizes *zoē*-centered egalitarianism. To do so, in the chapters that follow I attend lightly to the last century and a half of legal practices that frame children's rights and then pursue a hyper-relational, feminist alternative.

Monadic and singular forms of universal rights do not serve young people. Oswell (2013: 3) notes that, as they stand now, young people's "capacities to speak, act and become" are dependent on networks, assemblages or infra-structures involving "other persons and things in such ways as to endow [children] with powers, which they alone could neither use nor hold." These are problems that come from a complex set of policies and legal practices. The tension in Oswell's relational, postcolonial and postchild critique raises young people on the one hand as simultaneously and complexly dependent, independent, individual, aggregate, virtual and vital while on the other it views children's rights discourses and practices as representing young people in particular ways that do not necessarily advance their well-being. Although Oswell's perspective challenges the universality of children's rights, I want to argue that it is nonetheless possible to engage the spirit of universal child rights through spaces of convergence (that is, the ways that rights agendas transcend local constraints) without embracing the annihilation of difference that universality implies. This is not necessarily an easy or obvious task. For Oswell it begins by thinking of children as doings, where they are neither objects nor subjects of rights, nor are they a category of being or becoming. My thinking continues this with a locatable feminist politics that vehemently avoids the return to some regressive and romantic notion of place-based politics and *Gemeinschaft* nurturing. Today, young people are active in the construc-tion and the reimagining of their spaces so that they are seen and felt in families, societies and politics in ways that they were not seen or felt before. Young people are not simply beings or becomings, they are more significantly doings that have the potential to become and do something different, something yet unimaginable.

If the notion of the postchild and a feminist politics of location guides the theoretical thinking in this book then it traverses a portentous landmark in rights-based policy-making, and particularly what has transpired since the 1989 United Nations Convention on the Rights of the Child (UNCRC), which is arguably the most important and influential of any human rights agenda in 20th century. From the UNCRC, children's rights portend a politics of representation and a panoply of feelings that are very different now than they were at the beginning of the 20th century. This is not to suggest that prior to this adults did not feel for, cherish and love their children, nor does it suggest that children lived apolitical lives. Rather, through the late 18th century and the beginnings of industrialization in Europe and America, transformation in the ways children were attended to was joined by a focus on the ways they should and could be provided for and protected. This then evolves with the UNCRC into an idea that young people should participate in as much as possible that affects them. And yet despite all the legal and political efforts that are summarized in the pages which follow that might convince otherwise, many young people today are in a more dispossessed and precarious position than ever, while at the same time, ironically, their capacity 'to do' has intensified and the spaces in which they are able to do have proliferated. As Helena Pimlott-Wilson and Sarah Marie Hall (2017: 260) note, "examining the shifting nature of the economic landscape from the perspective of children, youth and families brings to the fore the effects of macro-economic changes for intimate, everyday geographies." They go on to call for an exploration of the interweaving of individual and collective responses and resistances. Given the current dispossession and precarity of young people, this book asks whether the UNCRC and the policy measures that followed are the right and appropriate focus for the well-being of young people in our postchild moment.

This book breaks from past work where I talk about young people's revolutionary proclivities (Aitken et al. 2011; Aitken 2014) to a more gradual reframing of everyday emotions and everyday politics. In the next chapter, I use the UNCRC as a reference point rather than a revolutionary break, because it is possible to consider that what is 'sensible' in the UNCRC may actually create a framework that forecloses upon young people's political acumen. I discuss the evolution of the idea of universal children's rights up to and including the aftermath of the UNCRC. I contrast the ways that these rights were prepared in light of the everyday politics of children's lives. Chapter 3 returns me to some of the philosophical ideas introduced in this beginning chapter and elaborates the notion of postchild relationalities more fully within the context of a radical understanding of play and erasure. I relate play/erasure to sustainable ethics in the sense that the latter comes about through the possibilities implicit in the former. Chapter 4 deals with a contemporary example of the erasure of young ethnic minority people's rights in Slovenia, which is followed in Chapter 5 with a focus on the ways these particular young people pushed back through the spaces and relations that were, and are, created and re-created through and with them. These two middle chapters provide an empirical fulcrum upon which the book rests. The real and painful erasure of young people in Slovenia after the country gained independence suggests a very interesting governmental and neoliberal complicity at play that in Chapter 5 I suggest enhances young people's capacities in good and important ways. Chapter 6 then extends the idea of young people's capacity to push back against erasure with examples from Chile, Romania, Brazil, and the USA. In each case, a unique and locatable feminist politics is suggested. For the idea of child rights to revive and have meaning in a postchild world, Chapter 7 contends with the complicit idea that rights are not necessarily attached to young bodies nor are they some kind of inalienable property associated with individual beings. Rather, rights stick with purpose to the lived worlds of children and young people through a sustainable ethics of care. It is from caring communities and radical ethics that young people are able to create and recreate their spaces and their lives so as to live life fully.

Notes

1 I am playing loosely here with the terms democratic and parliament. I prefer to think of democracy as an evolving process to the degree that we may not see in early democracies what we value today.
2 Indeed, it is worth noting that early on in this debate, Neil Postman (1982) was able to use Ariès' assumptions to deconstruct contemporary childhood out of existence. Unfortunately, Ariès' assumptions show up as a huge and distinguishing part of the UN Convention on the Rights of the Child (1989). In the pages that follow, and particularly in Chapter 2, I try to unpack this conundrum.

2   Locating young people's rights

This chapter comprises a broad discussion of children's rights; it engages the debate over the relations of those rights (both their accommodations and their tensions) to human rights discourses and practices in general, and it critiques the universality of those rights with a nod to different countries' contexts of child rights and advocacy development. This nod is brief but important because what I want to get to by the end of this book is an understanding of child rights that bring into play untapped possibilities for locatable community building. The nod in this chapter is part of an understanding of rights that moves away from a dislocated universalism to something locatable somewhere. That move is practical and political, and sits well with ideas of dissent from Jacques Rancière (2010, 2015), and it accords to a feminist politics of location (Grosz 2011) and posthuman sensibilities (Braidotti 2013; Murris 2016), which were introduced in the last chapter and will be raised again in Chapter 3. I then develop the contexts of dissent and locatable politics through the empirical chapters in the second half of the book, after which I am able to say something about sustainable ethics and locatable feminist politics. This second and somewhat ponderous chapter, however, focuses on the evolution of child rights discourses and charts, in an incomplete way, their progress from 17th century European Enlightenment through the rise of industrial capitalism in the 18th and 19th century, and the so-called neoliberal and globalization periods of the 20th century. The chapter weaves between early attempts to protect and make provisions for children, and efforts to create sustainable communities within which children could thrive. It looks at so-called 'child saving' movements and describes how those transformed into formal rights agendas. Those rights agendas are an important build up to the 1989 United Nations Convention on the Rights of the Child (UNCRC), arguably the most successful human rights treaty to date. I spend some time considering the politics and pitfalls up to the creation of the UNCRC. I also look at contemporaneous legal battles in the US that established rights and advocacy platforms in that country which, some argue, offset the need to ratify the UNCRC. I consider some of the limitations of legalistic and mechanistic, as well as universal and global proclamations that advocate for the well-being of young people. To the degree that neoliberalism today is offset by protectionism and populism at best, and fascism at worst, and globalization is questioned in the face of growing nationalist politics, the chapter ends with questions about where young people's rights are headed. This sets the stage for a discussion of the 21st century's Anthropocene, an era characterized by posthuman and postchild discourses, which is the topic of the largely theoretical Chapter 3.

Child rights redux

It is reasonable to assert that the modern idea of human rights in general, and children's rights in particular, originated in Western Europe and North America in the late 18th century coming hard on the heels of the industrial revolution and Enlightenment thinking. This does not mean that ideas about protecting children were not important elsewhere and at other times, but the formal idea of rights remains tied to particular societies and the emergence of particular forms of governance and economics. Nor does this mean, as argued problematically by Ariès (1962), that childhood and child rights as concepts in-and-of-themselves did not exist prior to the 18th century or that children were not active agents in their own subjectivities prior to modern times. Ariès' conceptualization of the invention and nature of childhood was focused on how they were represented (through portraiture for example) and who they were, rather than what they did. As suggested in the previous chapter, it is important today to re-conceptualize rights through a notion of what children do and create, as well as who they are, because young people's actions in societies and their contextualization in forms of governance elaborate spaces through which children's rights emerge. Therefore, in what follows, it is with a sensitivity about children as doings – of their intra-actions (Barad 2012) with themselves, others and things, and the assemblages of which they are a part (Oswell 2013) – that I approach their rights and everyday politics. With a consideration of what children do, of course, is the allied consideration of what rights do and it is to a consideration of this – as children's rights emerge from Western Europe and North America – that I turn to in this chapter.

The citizen-self

It is no coincidence that the rise of children's rights follows concerns for child welfare *outwith* (to use an old and apt Scottish phrase) parental oversight and *within* increasingly large-scale European and North American industrial capitalist institutions.[1] Prior to this time-period, under the auspices of feudalism and mercantilism, concerns for child-welfare, to the degree that they existed at all, were subservient to the authority of kith and kinship groups. Painting agrarian societies with inappropriately large brush strokes – from Mesopotamian Sumerians to Confucian Chinese and Cotton Mather's New England Puritans – it may be suggested that parents (most often the father) expected children's obedience and they did not 'spare the rod' in shaping responsibilities. In addition, more often than not, young people's responsibilities were to the family economy, with the family context (lauded by a father) over-shadowing concerns about anyone's individual welfare, children or adults. Of course, there were always important regional and cultural differences. Karen Wells (2015: 15) notes that in Islamic countries young children are the responsibility of mothers and only at seven years do they become the responsibility of fathers, and there is huge variability on when it is thought that young people are legally competent. Nonetheless, in his sweeping history of human-kind, Yuval Harari (2015) argues that there is no real rationality to our development as a species, but overcoming patriarchy seems to be an important stepping-stone for many societies and for Western European society in parti-cular, it resulted in Enlightenment thinking that presaged an unprecedented industrial/scientific revolution. The moral context of young people's lives, as seeding authority and obedience to a strict and patriarchal hierarchy, was part of the bedrock that Western Enlightenment thinkers sought to displace and this transformation, Harari points out, was initiated solely in Europe.

The new form of enlightened liberalism articulated by John Locke in the late 17th century presages the shift from child welfare to child rights during the industrial era. Locke's theory of the mind is the basis of modern conceptions of identity, particularly political identity, and the self as different from the family. Locke fundamentally disagreed with the contemporaneous theology that placed the family as the foundation of political life, and it is his attack on this form of patriarchy that prepares the way for individualistic liberal politics. The twin foundations of Locke's argument were that our primary desire is self-preservation, which translates materially into property; and our secondary desire is propagation, which shows up in familial life. This well-intended designation was Locke's attempt to upend Sir Robert Filmer's foundational *Patriarcha* (1680). David Foster (1994: 647) contends that for Locke to overcome Filmer's traditional notion of a divinely given law-of-the-father, he had to:

> make the family fully compatible with the primacy of the individual desire of self-preservation, and this has two main consequences. First, a concern for property comes to pervade all aspects of family life. Secondly, to make this possible, the relations of authority and obedience, which traditionally were considered the moral bedrock of the family, and which obligate parents and children to sacrifice property, liberty, and even life for the sake of family, must be loosened or undermined.

Locke's loosening of the connection between family and divine responsibility leads him to advocate rewards and friendship as more appropriate disciplinary tools for children than beatings and harsh language (Locke 1693: 106–8). From this, rather than binding children solely through filial connections, they become partners and relations; from here on in, tutelage rather than authority guides young people's lives. Locke pushed the idea of the citizen-self away from natural fealty (usually to a sovereign) to a construction based on actions and doings (in particularly, labor). As much as these ideas radically changed notions of property and rights, they perhaps left children in a more precarious position by taking away entitlement. Individuated (and propertied) citizenship was gained through the embodied labor that came with maturation, and so for Locke a young person was not yet entitled to the rights that were won when they became adults.

Locke argued further that children were neither good nor evil; against prevailing wisdom he noted that children did not need to be tamed by the rod because they were 'blank slates' upon which society could write its will via education (and through rewards, but also shaming). Locke famously claims in *Some Thoughts Concerning Education* (1693) that children are not rational but they have the capacity to learn reason. Children are on the way to becoming adults and the deployment of tutelage and education moves them along an appropriate trajectory to maturity. Locke's science and rationality posits children as initially neutral and gaining from paternal and societal optimism, but it also positions them as wholly dependent up until a certain age. Children in Locke's thinking were not rights-based citizens but were, rather, relationally tied and subservient to the productive proclivities of the family.

While setting up his famous social contract in the mid-18th century, Jean-Jacques Rousseau (1762) extends Locke's ideas to embrace children as essentially good, and society's role as maintaining a beneficent if not a rights-based stance towards that goodness, albeit in a paternalistic form. Through Locke, Rousseau and other Enlightenment thinkers in Western Europe and North America (notably David Hume in his *A Treatise on Human Nature* (1739/1955), whose work will be discussed more fully in the next chapter, and Thomas Paine in his famous *Rights of Man* (1791/1970)) evolved ideas of individual rights and shared equalities that grew rapidly and popularly with the expansion of science, capitalism, increased literacy and the Calvinist work ethic that characterized this unique and particular time and place (Harari 2015). From Enlightenment thinking, then, protecting and providing for the well-being of children was about protecting and providing for the future. Specific child welfare in the form of policies that focused on protection and provision did not take center stage until the abuses of the industrial capitalist system were realized, in a quite shocking way, by the growing middle-class of the 19th century.

Protection and provision

The shock of seeing children in precarious and dangerous industrial places fomented the first series of child-saving reforms. Although initially focused on a small minority of very poor children, the UK's *Factory Acts* were legislated to regulate labor as early as 1833, the same year that slavery was abolished in the British Empire. But it was not until the 20th century that children's rights in general were addressed and young people's rights to a life outside of work came to fruition. In 1833, Britain was still a rural nation with 80 percent of the population living in the countryside. Within the next two to three decades scientific agriculture turned farming into a capital rather than labor-intensive enterprise and unemployed farm workers and their families flocked to the cities and factory employment. By the middle of the 19th century over half the British population lived in towns and cities. When advancing science and technology created new types of work with the development of the factory system, it seemed perfectly natural to use children for work that adults could not do, such as crawling underneath machinery in motion or sitting in cramped spaces in coalmines to open and close ventilation doors. Nearly half the workers in a Manchester cotton mills in 1819 started at 10 years of age or younger, many working six days a week for between 14 and 16 hours a day. The 1833 *Factory Act* set the minimum age for employment at 9 years, while children between 9 and 12 years were allowed to work only 8 hours a day, and those between 13 and 17 years only 12 hours a day. The 1842 *Mines Act* restricted women and children to working above ground. *The Ten Hours Act* of 1847 restricted women and children under 17 years from working, and limited the workday to ten hours. These early acts were important precursors to notions of protecting and providing for children. If protection was about taking children out of harm's way, provision at this time took the form of looking after children's moral and spiritual welfare. *The Ten Hour Act*, for example, preceded other acts that required factory owners to set aside a certain amount of time for children's meals and education (including religious education), but none of these policies were rigorously enforced.

Many of the ideas behind the *Factory Acts* originated with the employment practices of Manchester factory owner Robert Owen who, in 1816, wrote *A New View of Society* based on his experience establishing a cotton mill in New Lanark, Scotland. By 1799 New Lanark was the biggest cotton mill in the UK and Owen began introducing a Utopian social experiment based upon enlightened humanitarian, social and educational ideas, revolving around the creation of self-supporting communities that nurtured children. I want to spend some time with Owen, because his ideas about community bonding fall in line with proposals for what is good about the contemporary posthuman moment from Braidotti (2013), Murris (2016), and others, with the exception that (as we shall see in a moment) Owen's ideas fall short of a practical implementation of a locatable feminist politics.

Braidotti (2013) argues that posthuman sensibilities come in part from environmental concerns (the posthuman as becoming earth), technological concerns (the posthuman as becoming machine) and a focus on new forms of community bonding. The idea of creating what Owen viewed as happiness through local environmental improvements, education for children using new pedagogies, and community cohesion, permeates large swathes of his *A New View of Society*. I am not in any way suggesting that Owen presaged the posthuman, but it is clear that his ideas about industrial life, and particularly the connection between production and reproduction, were different from what was evolving elsewhere as Vitruvian individuated rationality and reason. Owen did presage the Factory Acts by setting the minimum age for employees in his factory at 10 years, by stopping the practice of recruiting pauper apprentices, and by reducing the hours of a child's working day. An educational institute was established at the center of New Lanark with an enclosed space at the front set aside as a play area specifically to foster the creation of happiness.[2]

Owen's writing pushes against the inimical of the evolving capitalist system by attempting to dissolve the geographic separation of the public and the private that was increasingly developing almost everywhere in Europe and the USA in the early 19th century (see Aitken 2009: 41–7). Certainly, Owen's ideas coincide with romantic notions of connectivity, relationality and monism but they are also, in large part, about spatial rights and his foresight is that the separation of the public and the private is, in actuality, about control of those spheres. For Owen, maintaining spatial ambiguity between the private and public was about community bonding and, hence, for him, spatial cohesion was a coherent monism rather than an artificial dualism. It was about not separating too much the realms of production and reproduction, and giving workers and their children equal presence in an evolving community of work, relaxation, spirituality, play, education, politics, joy and happiness. Owen's *A New View of Society* was utopian in a good and important way, and it worked at a practical community level in New Lanark. His ambition to create a similar society in Indiana on a much larger scale failed hugely, however, because it did not realize a locatable feminist politics.

Location, here, is not about fixed location and, like Haraway's (1988) often misunderstood situatedness, it is not about social location along knowable axes of identity. Feminist locatable politics are not about local relations (although they can be), they are about specific connectivities and how body-minds intra-act with the world (Barad 2007: 471; 2012). Owen came part way to understanding how communities intra-act in the fulfilment of happiness, but he missed some important tensions between gender roles (i.e. actions) and gender relations (i.e. power) in the family (cf. Aitken 1998). He never completed his grand vision for Indiana's New Harmony which was purchased from its Lutheran owners in 1824 and was expanded and made operable for a while as a socialist community, in large part because his theoretical promises did not translate into practical application, and it is worth looking at why this was the case in terms of feminist politics and the practical contexts of rights. The impact of his progressive policy that both men and women vote on community issues (96 years before the 19th Amendment to the US Constitution gave women the vote and 40 years before the women's suffrage movement began) and his insistence that both boys and girls attend school were diminished by gender inequalities continuing in practice. Owen's farsightedness did not translate into the practical and emotional work of men and women. Married men and women were expected to work for the community but only women were expected to care for children, and although boys and girls went to school, the subjects that they were taught were gender specific (e.g. mathematics and Latin for the boys, and cooking and sewing for the girls). Female discontent was the primary part of the demise of New Harmony by 1828 (Kolmerten 1990). The issue of rights here, of course, eschews universalism in the face of practical equalities.

Outwith Owen's utopianism, by the late 19th century, a new spatial imaginary was growing in the rapidly industrializing and urbanizing Anglo-American world that had an important bearing on protecting children from hazardous work through rights agendas. Despite the effort of Owen and a few others (at least theoretically and ideologically if not practically) to protect and provide for children in self-supporting communities, the division of family labor was slowly reconstituted so as to place women and children in the so-called 'safe haven' of the domestic realm, separated from the work life of adult males (Aitken 1998: 46). Some feminists argue that the exclusion of women and children from the world of work had less to do with the Factory Acts in the early 19th century and more to do with the rise of organized labor and male dominated unions in the late 19th and early 20th centuries, which strove to protect the workplace and the 'family wage' as the exclusive domain of males (Mackenzie 1989). This spatial and social system reproduced an ideal labor force for the Global North's burgeoning economic system. From this it is reasonable to assert a series of undercutting questions, which run through this book, that relate to the extent to which the contemporary rights agendas for young people, including the UNCRC, fold into the world's predominant neoliberal, globalized economic and governance systems. Before I can consider this issue more fully, it is important to note that the contexts of children's rights are broader than those circumscribed by protection and provision; although on the surface they may well appear universally divorced from economics and politics, children's rights are clearly connected to the expansion of capitalism and imperial governance in insidious if not downright abusive ways. What do I mean by this and why is it important?

Countering this, and as a push against these abuses, children became the "first 'poor creatures' that the Victorian reform movements set out to rescue from the infernal cities that emerged in the wake of the first wave of industrialism" (Gleeson and Sipe 2006: 3). Gleeson and Sipe (2006) go on to suggest that the (primarily middle-class) social reform movements of the time were intent not just on saving children, but also on saving a capitalist system that seemed on the brink of self-destruction. Liz Gagen (2000a, 2000b) points out that the theoretical construction of the child that social reformers drew from made it necessary to display the instruction of healthy children in public, and particularly as seen in public parks, because they felt that children were a means towards larger social transformation (e.g. the perfectibility of Enlightenment principles). Jacob Riis' *How the Other Half Lives* (1890) famously portrayed the breakdown of the US social and economic system through the representation of urchins up to no good in grimy alleyways. At about the same time, G. Stanley Hall (1904, 1905) provided scientific evidence for the need to discipline and instruct children into society. He elaborated a neo-Lamarckian view that children pass through the same development stages as the species, and in adolescence, they are instinctively in tribalism. Hall argued that the desire of boys to form gangs at this stage can be re-orientated through team games in ways that competitiveness and cooperation might serve the state. By noting this, Gagen (2008) connects the beginnings of child development theory in the US with its imperial and capitalist expansion. Like children, peripheral colonial spheres were to be civilized with the imposition of governmental, economic and educational frames. The connection that Gagen elaborates focuses on domestic changes in the US propelled by the nascent discipline of development psychology's suggestion of what constitutes normal development, and the concomitant infantilization of so-called primitive cultures (i.e., that of children and colonies). Progressive attitudes to child development at home were coupled with the child welfare reform described above, and they were also exported from the US to its imperial protectorates through the establishment of schools, physical education programs and playgrounds abroad. Gagen notes that these two strands of development – one internal and one external – are all part of the same imperial project that is connected insidiously to economic expansion and the beginnings of a new world order and global economic restructuring that came to the fore after WWII.

In taking on the excesses and complicities in this new world order, Frantz Fanon (1967) recognized that critiques of both post-colonial imperial and anti-colonial nationalist ideologies were essential for 'decoloniality' (Maldonado-Torres 2016) in the Global South because when political independence is attained post-colonial and anti-colonial ideologies may converge to become a new mechanism for elites to exercise power over dissenters and marginalized people. In the second half of this book, I look at examples of marginalized youth and dissenters, and the contexts through which they push back. There is, however, another important critique of the continued colonization of young people's actions and bodies through the universal child rights that are created in and globalized from Western society, and it is to this that I now turn.

Universal rights

After WWII, welfare translated into rights-based concerns and expanded from a Western focus to become a global phenomenon predicated upon renewed contexts of national sovereignty, modern systems of governmentality and new forms of political expression (Oswell 2013: 234). The interleaving of governance, economics, science (and especially developmental psychology), and rights-based discourses is complicated, but it is nonetheless clear that the politics and economics of WWII hugely influenced the universality of the evolving discourses.

The universality of rights-based discourses emerged in large part from the sea of European refugees without nationality and legal status after WWII. As new nations emerged from the ashes of the war, thousands of people found themselves landless and stateless, as if they were excluded from humanity altogether (Arendt 1962: 297). With the post-WWII refugee crisis comes a specter of dispossession and precarity and, at the same time, there arises an idea of a global humanity to which rights must accrue. Children and young people are confounded through this framing because their status is seen as epiphenomenal and they are at best still viewed paternalistically through the Enlightenment legacy described earlier. While thinking of the development of child rights in this period, Oswell (2013: 235) links paternalism to the Greek notion of voice (*phōnē*), which was domesticated in the household (*oikos*) and politically marginalized from public spaces when compared to men's speech (*logos*) as part of the city state (*polis*) and public space (*agora*). As suggested by the late 19th and early 20th centuries reframing of women's and children's spaces by the US and UK Factory Acts, the domestic is disarticulated politically as childish, emotional, non-rational, and embodied in ways that are not logocentric. Paternalism at this time implies that rights accrued only to parents who then look after the best interests of their children, and from this perspective, the orphaned children amongst Europe's stateless WWII refugees presented an intractable problem that anticipates a larger perspective on child-welfare.[3]

Wells (2015: 24) traces the creation and proliferation of influential child charities and private philanthropy on behalf of children's welfare in Britain, France, Germany and the USA throughout the 19th and early 20th centuries, but the first major step on behalf of the global welfare of children was the creation of the United Nations International Children's Emergency Fund (UNICEF) in 1946, to provide emergency food and healthcare to orphans and refugee children. UNICEF was set up to mediate crises involving young people, and it was only later that its mission became focused on the protection of children's rights. Before that could happen, a broader discussion of rights was required. In 1948, the Universal Declaration of Human Rights was established as part of the United Nations Charter. It is important to understand that although the declaration has roots in the carnage of WWII it was constructed largely at the behest of the US, foreshadowing the beginnings of the Cold War. In practice, there are several reasons why the declaration proved a relatively ineffectual political tool. First, it merely extends Enlightenment thought to forms of global and universal rationality that are simply untenable in a diverse world. Thinking back to my earlier points on Enlightenment thinking, Paine's (1791) universal 'man' simply does not exist, and it is hugely problematic to consider the rights of women and children under Locke's (1693) paternalism. The individuated context of Locke's liberalism does not easily encompass difference and diversity, and attempts to do so more often than not end with repression and domination, under the shadow of the Vitruvian man.

Moreover, and importantly, the second half of the 20th century witnessed growing academic and political debate on rights claims based on locality, embeddedness, and cultural history. This began, argues Harvey (2000: 87) with the American Anthropological Association (AAA) contesting the specificity of the 1948 UN Declaration with the idea that respect for the cultures of differing human groups was equally as important as the rights of the personal/universal individual. The AAA went on to draw attention to the fact that the UN Declaration was narrowly conceived in terms of values prevalent in the countries of Western Europe and North America; it cautions that "[i]n the history of Western Europe and America ... economic expansion, control or armaments, and an evangelical religious tradition have translated the recognition of cultural differences into a summons to action" (quoted in Harvey 2000: 87). Put simply, the culture of Anglo-America is the culture of capitalism and that culture pushes homogeneity against difference, often with suppression and controls in the name of universal (liberal and neo-liberal) ideals about how markets and governments should work for the benefit of all. The idea of universal market ideals, then, is implicated in the evolution of the idea of universal human rights.

This implication notwithstanding there is tension and confusion, conceptually and in practice, about how human rights agendas deploy, first, in the face of rights agendas for so-called vulnerable groups (e.g. refugees, migrants, women, and children) and, second, with regard to civil and political rights, on the one hand, and economic, social and cultural rights on the other. The universality of human rights is contested practically and geographically from the latter perspective, for example, with Eastern European rights activists pushing civil and political rights agendas most virulently whereas Latin American rights activists tend to focus on economic, social and cultural rights (Reynaert et al. 2015). Harvey (2000) points out that globalization and the rise of transnational forms of capital and neoliberal forms of governance after WWII make it difficult to sustain a separation between these rights forms. There are clearly important locational differences that the empirical chapters in this book will highlight, but to the degree that neoliberal agendas push uncertainty onto the shoulders of workers, and responsibilities for childcare and education onto parental shoulders, it becomes increasingly difficult to separate civil and economic rights. Tension between so-called vulnerable groups in a neoliberal era when linked to the struggle to make transnational institutions accountable suggests that the question of economic rights must bear heavily on any restatement of universal rights.

The 1959 UN Declaration of the Rights of the Child presaged the 1989 UN Convention on the Rights of the Child (UNCRC), and has its basis in a 1924 League of Nations document that guided the 1948 Universal Declaration of Human Rights. Like the 1948 document, the 1959 Declaration did not have force of law, but merely provided guidelines. To the degree that the interests of children emerge only in relation to community and societal interests, it is worth noting that the 1959 declaration added, channeling Owen, a commitment to a "happy childhood that would serve both the individual and society, and that legal protections should allow the full physical, mental and spiritual development of the child" (Stearns 2017: 15). To understand the ways this latter declaration foreshadowed the impact of the UNCRC and the importance of legal and political precedents, in what follows I contextualize legal changes to child advocacy and rights. I spend time with these legal briefs because the changes they wrought set part of the stage for 1989's UNCRC, and they aid understanding of the Cold War tension that pushed the emergence of a global child rights agenda squarely onto the European stage.

Legal precedents to global children's rights

The advent of the Civil Rights movement in the US generated an important expansion of efforts in the area of children's rights and child law from the 1960s onward. Various US civil rights leaders such as Richard Farson sought to include children in basic rights claims, arguing that they should be fully free to express themselves, associate with whomever they pleased, and even leave their families if they chose (Stearns 2017). None of this found its way into US law although through Farson, the Civil Rights movement spawned the Youth Liberation Movement. Wells (2015: 15) points out that "the field of child law ... [focusing] on the legal competence of young people and the necessity of separate legal procedures for dealings with minors date[s] back to at least the 16th century," but in the1960s, at the height of the Vietnam War as thousands of young American men died, it seemed to herald a particular urgency in the US.

Perhaps the most important US-inspired precursor for the UNCRC was a series of court decisions that challenged the adult focus of the US's 14th Amendment. The 14th Amendment was adopted in 1868 as part of the so-called 'Reconstruction Amendments' following the American Civil War; it addressed citizenship rights and equal protection of the laws, and is one of the most litigated constitutional amendments. The 14th Amendment formed a basis for later cases that focused explicitly on spatial (e.g. land ownership rights and school segregation) and identity (e.g. reproductive rights and same-sex marriage) issues. It is worth looking at *In re Gault* (1967), the first landmark US court decision that changed the amendment from an adult- to a child-rights focus for three reasons: first, it ties into similar tensions between adult and child rights that ensued from the UNCRC, second, it raises interesting issues of *habeas corpus* and third, it suggests that the severity of punishments for specific crimes links judgements and judicial institutions in important ways.

Jerry Gault, a 15-year-old from Gila County, Arizona, was accused of making an indecent phone call to a neighbor, Mrs. Cook, on June 8, 1964. After a complaint was filed by Mrs. Cook, Gault and a friend, Ronald Lewis, who was hanging out with him at the time of the phone call, were arrested and sent to a juvenile detention facility. The arresting officer did not attempt to contact Gault's parents, who were at work at the time. When she got home, Gault's mother sent Jerry's brother to look for him but they did not learn of the arrest until they were contacted by Ronald Lewis' parents. When she got in touch with the authorities, Mrs. Gault learned that Jerry was arraigned and scheduled for Juvenile Court the next day. The next day's hearing was informal; Mrs. Cook was not present, nobody was sworn in and no transcript was taken so it is unknown what Jerry admitted to, if anything. After the hearing, Jerry was taken back to the juvenile detention facility for two days and then released to await another hearing, scheduled for June 15, 1964. Mrs. Cook again was not present at the second hearing despite Mrs. Gault's request that she be there to determine which boy, Richard or Ronald, "had done the talking, the dirty talking over the phone." At this hearing, probation officers filed a report listing the charge as lewd phone calls but neither Jerry nor his parents were made aware of this report.[4] At the conclusion of the hearing, the judge remanded Jerry to State Industrial School for six years, until he turned 21. An adult charged with the same crime at that time would have received a maximum sentence of a $50 fine and two months in jail. Gault's parents filed a petition for a writ of *habeas corpus*, which was dismissed by the Superior Court of Arizona and the Arizona Supreme Court. They then brought the issue to the Supreme Court of the United States, which agreed to hear the case to determine the procedural due process rights of a juvenile criminal defendant where there is a possibility of incarceration. In a unanimous decision, the Supreme Court overruled the denial of counsel to indigent defendants, and noted that if Jerry Gault had been 18 he would have been afforded the procedural safeguards of an adult. In its deliberations, the Court closely examined the juvenile court processes and determined that, while there are legitimate reasons for treating juveniles and adults differently, young people facing an adjudication of delinquency and incarceration are entitled to the procedural safeguards designated by the Due Process Clause of the 14th Amendment.

The details of this case are important because they raise questions of the suitability of courts to appropriately deal with a young person's rights when establishing the truth of alleged offenses by minors while also adjudicating a suitable way forward (i.e. a punishment). The drama and fumbling of the Gault case questions the fundamental distinctions between the rights that children have under law and their moral rights (Archard 2004). If legal rights are determined judicially and factually through court records, then the Gault case also questions the ability of adult courts and legal systems to deal with minors. If the Gault case is about distinguishing the rights of adults and children, and questioning the ability of an adult-based judicial system to adjudicate a way forward for child offenders, a second important US legal battle presages the UNCRC focus on children's political voices.

Young people's free speech came to the fore in *Tinker vs De Moines* (1968), where the US Supreme Court affirmed that school-aged children are 'persons' possessed of the same rights as adult citizens under the 1st Amendment of the US Constitution, which affirmed, amongst other things, the right to freedom of speech and peaceably assemble. Mary Beth Tinker and her brother John of Des Moines, Iowa, were 13 and 15 years of age respectively when they decided to hold a protest against the Vietnam War by wearing black armbands during the 1965 Christmas season and to support the holiday truce called for by Senator Robert F. Kennedy. Mary Beth and John were joined by their younger siblings Hope (11 years old) and Paul (8 years old), along with their friend Christopher Eckhardt (16 years old). The Tinker children's actions were condoned and supported by their parents who were also anti-war activists. Aware of the plan, the principals of the Des Moines school district met and adopted a policy that any student wearing an armband would be asked to remove it and if they refused, he or she would be suspended. Mary Beth Tinker and Christopher Eckhardt were suspended from school for wearing the armbands on December 16 and John Tinker was suspended for doing the same on the following day, while the two youngest participants were not punished. Mary Beth, Christopher, and John were suspended from school until January 1, 1966, when their protest had been scheduled to end. The parents of the children filed suit with the US District Court when they were urged to do so by the Iowa Civil Liberties Union, and the American Civil Liberties Union agreed to help. The lower court upheld the schoolboard's decision, and a tied vote at the 8th Circuit Court of Appeals pushed the suit up to the US Supreme Court, where it was argued on November 8, 1968. In a vote of 7–2, the Supreme Court ruled in favor of the Tinker and Eckhardt parents, emphasizing that students have First Amendment rights, noting that "[i]t can hardly be argued that either students or teachers shed their constitutional rights to freedom of speech at expression at the schoolhouse gate." The ruling was controversial, and it was further noted that free speech could be disallowed if the manner in which it was voiced was disruptive to the running and disciplining of the school. The latter, of course, becomes a matter of significant debate when perceived tensions supersede seeming free speech. For young people on a schoolyard, the idea of belonging to a particular place necessarily creates the potential for disruptive interactions. This is especially true in conceptualizing the schoolyard in terms of its many territories and micro-geographies. When a particular group occupies a certain place in the schoolyard they may set the conditions for use of that space that is protected by the 1st Amendment only to the extent of state and school regulations (Aitken and Colley 2011).

Tinker vs Des Moines remains a viable and well-cited Court precedent. Mary Beth Tinker went on to become a noted activist for youth voices, but later Court rulings limited the free speech rights of students. In *Bethel School District vs Fraser No. 403* (1986), for example, the Supreme Court ruled that a high school student's sexual innuendo laden speech during a student assembly was not protected under the 1st Amendment, creating an exception for 'indecent' speech. In *Hazelwood School District vs Kuhlmeier 484 US 260* (1988) the supreme court ruled that schools have the right to regulate the content of school-sponsored newspapers if they contravene legitimate educational contexts without violating students' 1st Amendment rights. Further still, in *Morse v. Frederick, 551 U.S. 393* (2007) – when Principal Deborah Morse suspended student Joseph Frederick for displaying a banner spelling out 'Bong Hits 4 Jesus' across the street from Juneau-Douglas High School during a school relay race – the Court held that schools may, consistent with the 1ˢᵗ Amendment, restrict student speech at a school-sponsored event, even those events occurring off school grounds, when that speech is reasonably viewed as promoting illegal drug use.

I spend some time detailing these US court cases for several reasons. First, the US is often judged harshly for not ratifying the UNCRC and although President Obama was outraged that this had yet to happen, by the end of his presidency the US was the only country in the world left out of the UNCRC, when Somalia and South Sudan became signatories. One of the most cited reasons for the US position on the UNCRC is the rise of family values since the 1990s, and in particular the lobby-power of parent groups, who argue that children's rights detract from parent rights and thus undermine family values. Second, the plethora of court cases in the US from the 1960s onwards highlight some important decisions that distinguish child rights from adult rights to the degree that many US policy-makers felt that child rights were best handled nationally. Third, and relatedly, the US for the most part remains outside of international human rights efforts. For example, it is one of only seven countries that has failed to ratify the Convention on the Elimination of All Forms of Discrimination against Women (CEDAW). Although it may be argued that the failure of the US to join with other nations in taking on international human rights legal obligations has undercut its credibility in promoting respect for human rights around the world, others argue that taking on these issues would diminish the US's legal autonomy (Human Rights Watch 2009) Finally, to the degree that the UNCRC arose in the midst of the Cold War, there are some interesting politics, to be discussed on p. 30–32, that pushed discussions away from the US and towards Eastern Europe.

Before returning to the UNCRC and its construction during the Cold War, it is worth visiting one more piece of legislation – a somewhat unique practice of dealing with children's legal rights – that arose elsewhere at about the same time the US was beginning to deal with its modern distinctions between child and adult rights. I do so with a particular intent: I want to set up a slight tension – by no means binary and hopefully creative – between those who create policy and law and the actions and activism of young people. I do so with the next example to make clear that there are important tensions, which lead to positive outcomes, and that the state is an important contributor to those outcomes. The theoretical arguments for outcomes of this kind are discussed in Chapter 3, and the important and complex relations between young people, activism and policy are part of the empirical contexts of Chapters 4, 5 and 6.

Listening to young people outwith the judicial system

According to David Archard (2004), the most far reaching and progressive act on behalf of children's legal rights was the 1969 Scottish Social Work Act.[5] The Act set up the *Children's Hearings System*, and is, perhaps, the exact opposite of how US child right laws proceeded after the *In re Gault* (1967) case. Specifically, it cast doubt on the penchant for creating judgements about child rights through the same judicial institutions that are used to adjudicate adult rights, whether or not those rights are deemed different. Scottish society, through the *Children's Hearings System*, established that young people are different from adults and, as such, there are better ways of proceeding with a child who has allegedly committed an offence than the requirement to appear before a court of law, which also determines the punishment. The *Hearings System* is a quasi-judicial instrument for dealing with children who commit offences and children who are the victims of those offences, but it is also a harbinger of rights that are tailored around the contexts of offences and derived from consensus. From this perspective, child rights are elaborated from local contexts that eschew universal proclamations: the conversation is kept going because each case is deemed unique. Children's hearings took over from the courts most of the responsibility for dealing with children and young people under 16, and in some cases under 18, who commit offences or who are in need of care and protection. Archard (2004: 130) points out that the Scottish Hearings System is "unique and extra-ordinary" because it establishes the children in question (both perpetrators and victims) as "children in trouble," and it deals with them outside of the court system. By so doing it keeps the conversation going and eschews the enervating totality of universal proclamations (see discussion below). The Scottish system aims to ensure the safety and wellbeing of vulnerable children and young people through a decision-making Children's Panel, comprising trained and appointed lay-persons. This panel comprises people with the knowledge and experience necessary to consider children's problems, and was a model on which none of the then current systems of juvenile justice in the US or Europe were based. The model of US and UK juvenile courts were deemed unsuitable because they combined the characteristics of a criminal court with those of a treatment agency; the *Hearings System* separates the issue of determining truth from the outcome of the case and establishes a basis for listening to children. Children and young people attending hearings have a range of legal rights to ensure preparation and participation, including requesting a prehearing or deferral, or bringing a representative with them to the hearing (e.g. friend, family member, lawyer), and rights to appeal. A report celebrating 45 years of the *Hearings System* using qualitative methods on a representative sample of parents and children noted that panel members "were … friendly, professional, courteous and empathetic" and the Hearing System created "an informal atmosphere" (Homes et al. 2014: 22). The decision of the hearing was "… rarely a surprise and participants were generally either happy with it or at least understood the hearing's reasoning. Participants usually felt their thoughts had been taken in to account" (Homes et al. 2014: 55).

Suspicions about claims for a definite and static, mechanistic version of justice based on logic and reason hinge on its basis in Enlightenment thought (Flax 1990, 1993). The *Hearings System* offers an alternative discourse of discussion and negotiation between panels and children in trouble that seeks to neither marginalize nor prioritize any one point of view; that is, it does not prioritize the voices of perpetrators or victims, nor that of adults. This relational view is an important push against traditional notions of justice. Traditional versions of justice typically involve either some hierarchical and arbitrary valuation of difference or, in the case of universal rights, some uniform treatment of difference that, while appearing more equitable, disguises the real and ongoing forms of domination that exist in the construction of child rights, and stops ongoing conversation before it begins. The Scottish *Child Hearings System* is fluid, and conceives justice as a process made up of interrelated and interdependent practices of communication. Another strength of the fluidity of the *Hearings System* is that it is able to adapt to changing social and political climates. The fundamental principles upon which it is based have been maintained – but processes have been changed in light of international conventions, including the specific rights for children contained in the UNCRC.

In the 1960s through the 1980s, different countries came up with different ways of particularizing children's rights. The US and the Scottish examples in this section suggest very different approaches to child rights as they relate to juvenile offenders. In the former, specific court cases are established as bench marks, which are established through a rational and mechanistic legal process and become part of the rule of law. In the latter, decisions about children's fates are made outwith the legal system in a process that is about hearing, discussion and consensus rather than adjudication. I will return to this local, relational perspective from a more theoretical angle in the next chapter, but at this point it important to return to the establishment of the most ratified, most discussed, and arguably the most important human rights treaty in the 20th century. In what follows, I discuss the global political machinations that brought the UNCRC to fruition, its influence to date, and its limitations.

The United Nations Convention on the Rights of the Child (UNCRC)

Although it is important to recognize the progress made on children's rights prior to the UNCRC, particularly in the US and the UK, it is nonetheless clear that some kind of sea-change occurred in 1989 and the impact was felt broadly and continues today. It is also important to note that although this universal treatment of children's rights is critiqued for its Western biases, it emanated from Eastern Europe. Be that as it may, the UNCRC instituted a sea-change not just in how we think about children, but also in how we think about rights.[6] First, it established a platform for an unprecedented global discussion on children and rights. Second, and perhaps more importantly in terms of a transformation in how we think, it created a distinction between human rights and children's rights.[7] The UNCRC also resulted in activists for children's rights moving in a direction that differed from those focusing on human rights, by converging on young people's participation in governance processes that were heretofore absent, as well as providing for basic needs and protection from harm. Young people were not so much to be represented by adults, but to engage with adults and participate in processes that had bearing on their well-being. Prior to the UNCRC, with the exception of some of the US cases described above, minors were often seen as epiphenomena, coasting (and more often dismissed) on the coat-tails of parents' or guardians' rights. Alternatively, they were seen as passive recipients of welfare, and dependent on adults for their well-being. With the post WWII refugee crisis in Europe and Asia, and the independence of African and Latin America colonies in the 1940s, 50s and 60s, came a recognition of the number of minors embroiled in larger political processes, buffeted to and fro in a sea of political change, and social and economic restructuring. There is an important geography of rights here: the recognition of the need to look out for young people in interconnected spaces and processes, and in a globalized world that transcends national boundaries, really only came in the mid-20th century, pushed by the post WWII reorganization and rebuilding of large parts of Europe and Asia, and the withdrawal of colonial powers from Africa and Latin America.

To the extent that the UNCRC provides a broad platform for discussion of children's rights and is adopted by almost all countries suggests that there is no doubt about its importance and its global reach. That the transformation is monumental in its scope and adoption is beyond doubt, but what precisely does this mean? Many commentators argue that the UNCRC continues a long tradition of colonial and imperial discourse with a specific focus on European definitions and understandings of childhood. Moreover, it is not just the export of this worldview (and legal processes) that is colonial, it is concern that what is elaborated continues as a worldview that colonizes children and young people. Specifically, with the UNCRC's focus on three Ps, protection and provision are seen as paternalistic, and participation (elaborated through UNCRC Article 12) evokes a quasi-Lockean view of the monadic individual. It is not inappropriate, then, to label the construction of the UNCRC as influenced by Western Enlightenment thinking. This thinking comes from ideas discussed previously in this chapter, but it is also important to consider the larger geo-political issues that shaped the UNCRC's legislative trajectory, and it is to that that I now turn.

Eastern European origins and Cold War deviations

On November 20, 1989, the General Assembly of the United Nations adopted the UNCRC, framed in 54 articles.[8] The convention is indebted to a history of earlier international declarations including, as noted, the League of Nations Declaration of the Rights of the Child in 1924 and the 1959 eponymous United Nations Declaration. Significant, also, is the founding in 1946 of the United Nations International Children's Emergency Fund (UNICEF). This European delineation is shaped in a particular social and historical context of power and exploitation. To the extent that this was a long struggle for the recognition of children as fully fledged human beings it is also couched in a uniquely Eastern European context of exploitation. The submission of a draft convention on children's rights by Poland to the UN Commission on Human Rights in 1978 was in large part to make clear, first, that human rights in general was not a monopoly of Western states and, second, that Poland under Nazi occupation during WWII gave that country a singular perspective on children's rights. The early legislative history of the UNCRC, then, highlights Polish sensitivities to the plight of children:

> During the First World War and even more so during the Second World War, children in Poland experienced suffering that is hard to describe. It was caused by the wartime operations taking place on Polish territory. As a result, many children starved, were deprived of basic health care and of access to education, and were forced to perform difficult and excessive work. During the Second World War children and their parents were massively displaced from their homes and many were taken from their families in order to undergo Nazi indoctrination. Children of Jewish and Gypsy origin were victims of extermination.
>
> (Lopatka 2007: xxxvii)

Poland was the only country where the Nazis set up a concentration camp for children. Between the two world wars, Polish educator, philosopher and medical doctor, Janusz Korczak developed a contemporary notion of childhood based upon the conviction that the child is an autonomous person. His disappointment that the 1924 Declaration on Child Rights by the League Nations did not go far enough prompted him to write an essay where he declared that "[c]hildren are not the people of tomorrow, but are people of today. They have a right to be taken seriously, and to be treated with tenderness and respect. They should be allowed to grow into whoever they were meant to be" (Korczak 1929: 7).

Locke's radical view of the relationship of children to their parents and to the polity (a rejection, as noted earlier, of a feudal and authoritarian hierarchy from royal authority downwards) focused on the fact that young people in time are intended to occupy an equal and independent status as mature, rational beings. For Locke, the state also possessed limited jurisdiction over children, for its duties should conform to the obligations imposed by natural human rights that were inviolable and inalienable. Korczak moved these ideas further by ignoring Locke's insistence on maturity (a nebulous term at best) and advocating self-government through a Children's Parliament and a Court of Peers in the orphanages that he established. At the court, any child could accuse persons (other children and adult staff) of behaving inappropriately. For Korczak, children's rights were a foundational democratic issue if they were also based on a form of relationality that encompassed at its core communication between children, and between children and adults. Not unlike the *Scottish Hearing System* that came later, the Court of Peers was a way to create communicative rather than instrumental justice (Hartman 2009: 17).

This notwithstanding, Korczak's notion of relational and communicative democracy did not translate fully into the UNCRC, which continued during construction with a Western developmental notion of children that was supported through family, schooling and the protective welfare states that evolved after WWII and under the shadow of the Cold War.[9]

The Polish draft convention was sent to UN member states for consideration in 1987. Some were in favor of the convention including civil and political rights as well as economic, social and cultural rights, while maintaining an appropriate balance between them. The USA and, in particular, the administration of President Carter under a Cold War cloud, suggested that the draft focused too heavily on social, cultural and economic rights while not pushing children's political and civil rights. Based on the suggestions from UN member states as well as international and non-governmental organizations, the Polish Government prepared a second, amended version of the draft convention. As the draft convention went to UN Working Groups tensions arose from the USA and the USSR, with both parties using tactics of obstruction. The rights of specific groups such as orphans, refugees and children with disabilities were raised as paramount; controversial alternative proposals were submitted and then retracted when consensus was not reached. Time pressures between the UN's child rights working group and other working groups arose. Projects of particular importance to some Western states, such as the Convention against Torture, for example, were pushed ahead of the work on children's rights. With the outcome of the UNCRC seemingly less certain, UNICEF – which had not heretofore been involved – contributed financially and with personnel support so that the draft could be finalized before 1989, the 13th anniversary of the adoption of the Declaration of the Rights of the Child and the 10th anniversary of the International Year of the Child. UNICEF is recorded as showing a "total lack of initial interest in the exercise" and did not send strong delegations to the Working Group until 1986, by which time it had become clear that the text of the UNCRC would be put before the UN General Assembly for adoption (Cantwell 1992).

By early 1989, the USA and USSR were no longer obstructing the draft convention, with the Carter administration espousing satisfaction with the insertion of political and civil rights. This notwithstanding, the tension between political and civil rights on the one hand, and social, cultural and economic rights on the other has played out with the implementation of the UNCRC over subsequent years.

Childhood, universal rights, and neoliberal logics

As noted above, Western constructions of rights are embedded in the Enlightenment thinking of Locke, Paine and others, and this thinking embeds the *UN Charter on Human Rights*. The notion of children's rights is equally influenced by these Western traditions, and even although it began its life in Eastern Europe the basis of the UNCRC is a categorization of children as monadic and part of a coherent political group ('beings' rather than 'becomings'). Moreover, for the UNCRC, childhood as a state is defined by young people's age, physiography, needs and desires irrespective of class, gender, race and ethnicity. In the 1990s there was an eliding of these intersectional distinctions in favor of a universal notion of child rights. So it is reasonable to critique the individualistic premises of the UNCRC and its all-or-nothing character, which for Archard (2004: 113) exacerbates the modern tendency to keep the worlds of adulthood and childhood separate. He goes on to suggest that the UNCRC is morally impoverished because it neglects an alternative ethical view of the world, in which the affectionate, caring interdependence that ideally characterizes the parent–youth relationship assumes an exemplary significance. Coupled with a neo-liberal, globalized discourse, in setting up the child as a monadic self, the UNCRC creates childhood as flexible in the face of global change. Perhaps the biggest problem with the UNCRC is that as an agenda that touts universalism by definition it stops the conversation, which is a characteristic that falls in step with other universal rights discourses.

Ironically, these limitations of the UNCRC helped set the stage for copious academic discussions, which propelled in large part what is now known as the 'new sociology of childhood' (cf. Jenks 1996; James et al. 1998). The discussion not only provided insights about the ways that the voices of young people could be incorporated into policy decisions, it also raised larger issues of what precisely constituted the child and childhood, young people and adolescence, and it suggested questions of what rights might accrue to different life stages. These arguments held sway until poststructural and posthuman perspectives challenged the social construction and demarcation of categories. Coming from a relational perspective, Oswell (2013: 14–17), for example, articulates six problems with sociologists' overemphasis on the category 'childhood': (i) understanding childhood as a category or, worse, a focus of political identity, often fixates on the category having power and agency in and of itself, and less on the assemblages of processes, actors and agencies that do the work of classifying; (ii) categorical thinking assumes neat boundaries and a logic of identification (e.g. minors are under 18 years of age) that elides messiness; (iii) the boundaries prescribes a box within which not all children are comfortable; (iv) categories often presume to be filled by individual children, and so the new sociology of childhood repeats and reframes a notion of individual agency, which assumes children are always agentic or have the capacity for agency; (v) the category 'childhood' prioritizes language or cognitive schema over media, technologies, affective relations, cultural forms, and the myriad of ways children come together; and (vi) to the degree that the category 'childhood' does not define children then its use is as a problematic point of reference, mobilized only for particular social and political reasons. For example, the provision in the UNCRC that children should be involved in discussions about policies that affect them results in some far reaching and unanticipated concerns about the remit and reach of the UNCRC, the implications of children's participation, and how young voices are incorporated in political decision-making.

Examples of so-called participation and incorporation run the gamut from tokenism to outright exploitation. Even young people who are activists and who are politically perceptive may be represented only symbolically at national forums and international conventions, and sometimes the representations are in ways that are of little use in improving their life situations (Kjörholt 2008). Perhaps more disturbing is the tendency to groom young people for political office through national policies created with the ratification of the UNCRC. Kallio and Häkli (2013), for example, note that the children's politics in official Finnish settings are almost always different from young people's politics in everyday settings.

Radical critiques of the UNCRC

By the early 2000s the voices of researchers and activists concerned about social/spatial justice and contexts of difference (including locational difference) pushed against the notion of universal child rights and the categorical ideas from the new sociology of childhood. To the degree that the UNCRC is the most successful of all the rights-based agendas since WWII in terms of its scope and adoption, Fernando (2001: 8) points out that there was nonetheless a widening gap between expectations and achievements for human rights as a whole and this has ramifications for "how comfortably children's rights fit with our convictions about social justice." As Kallio's Finnish example suggests, its importance as the impetus for a series of academic studies, policy debates and practices that created new contexts for understanding children and their place in the world is also clear, but therein lies a crucial geo-political and social justice transformation with which we (adults and governments) are still coming to terms.

To the degree that issues and social justice questions about categories such as 'childhood' or 'the child' are of pressing concern, they are addressed, delimited, and contested around the world in interesting and provocative ways. And where they are not addressed, larger issues of problematic social, economic and spatial structures and political foreclosures are highlighted as broader human rights, which map onto children and young people in convoluted and problematics ways. Fernando (2001) points out that the key issue of universal child rights is fundamentally an issue of distributive justice. When the UNCRC sets up a stage for the child as competent social actor, it misses the detailed need for a comprehensive plan for a new and more just economic system and, perhaps, provides instead a model for the child as a neoliberal subject. As Fernando (2001: 8) notes:

> Although no signatories to the [UN]CRC, international donor agencies, or child rights advocates would deny that the present system of distributive justice is mainly responsible for the vulnerability of children, they ultimately shy away from a serious dialogue about alternatives to it.

He goes on to note that academics are equally at fault for not offering a critical theorizing and a political consciousness that prescribes reflective action. The discourse on children's rights as elaborated by the UNCRC and taken up by ratifying governments is becoming technocratic to the extent that it no longer addresses the issue of power relations (Fernando 2001: 12). Just as mainstreaming gender peripheralized women's needs in development studies (Brun et al. 2014), so the mainstreaming of children's rights marginalizes the real issues behind children's needs: "isolating children's rights from issues of class, race and gender ... has become a convenient means for avoiding direct engagement with the political and economic realities of the emerging global economy" (Fernando 2001: 12). In the context of other human rights (women, refugees, etc.), the hope that Fernando lays out resides with the universal entreaty that children inspire for intercultural consensus, but he points out that it is impossible to isolate children's rights from their embeddedness in class, race and gender relations experienced through their relationships with other children and adults.

In what ways does the UNCRC and particularly what followed it, systematically contrive an imaginary of the child that is politically, socially, economically, ideologically, and scientifically a Euro-centric construction? Before that question can be addressed it is worth considering what Erica Burman (1996) calls a move from generalization to naturalization in universal child rights proclamations, and the ways these not only fit a Eurocentric but also a neoliberal projection. In a wide ranging discussion of the UNCRC's impact, she notes the successes and limitations of international policies and programs as they move from the global to the local to the globalized. Burman's perspective is particularly germane to the discussion here because of the ways she ties it to child development and other issues that I cover above (see also Burman 1994 and Gagen 2008). Burman's (1996: 47) contribution relates to how tensions between the local, global, and globalized conceptions of childhood are "played out within, and maintained by, the version of child psychology subscribed to by UN policies and programs for children." Like Oswell, she is quick to point out that at the level of policy and the practice of rights legislation it is important to not only be aware of normalized definitions of what it is to be a child, but also how those connect to larger ideologies. Burman's standpoint is that the Western ideology of liberal individualism which gave rise to rights legislation is distinct, decontextualized and conflicting in ways that map uneasily onto Global South settings. She questions whether general statements about children and childhood can be made and then put into practice. To the degree that claims to generalities are fraught with ideologically abstractions and vagueness, Burman (1996: 47) points out that practitioners' interpretations are case specific and culturally sensitive: in short, concepts of rights and childhood are necessarily local.

To cite a prominent example, on July 17, 2014, Bolivia enacted a law allowing children as young as 10 years old to work in paid employment. After decades of proclamations from institutions like the UN and the International

Labor Organization (ILO), the new law flies in the face of ideas that young lives should be dedicated exclusively to education and play, and that they should be protected from the seeming harsh realities of the world of work. Under Bolivia's new law, 10 year olds can work if they are self-employed and if they are also attending school. The legislation also sets 12 years as the age when a child can work under contract, assuming that they have permission from their parents or guardians and continue in school. These new criteria expand Bolivia's 'Code for Children and Adolescents', which previously followed ILO recommendations of 14 years of age as the minimum for child labor. The new law was immediately criticized as a step backwards in its attempt to bring Bolivia out of poverty (Krishnan 2014). But in acknowledging that a large swathe of Bolivian children work anyway, the law was also an attempt to ensure better provisions for young people's physical and mental health. To that end, the law put in place restrictions on employers; it enacted severe consequences for violence against children, and required voluntary consent from child laborers and their guardians, and mandated permission from a public ombudsman before children could work (Thompson 2014). The law ensured an outright ban on children working in mining, lumber and other hazardous industries. To the extent that the law is the first clear counter to the 'abolitionist' sentiments of the ILO and the UN it perhaps reflects more clearly the complex nature of child labor and establishes a more nuanced and effective regulatory strategy. This complex nature is contextualized in systemic poverty that cannot be eradicated by simply removing children from productive work. The Bolivian labor law begins from the premise that children work because they are poor and that, until their poverty is overcome, they are better served by having their work brought out of the shadows of illegality. This involves having child labor legalized and regulated, and ensures that child workers receive the same protections and wages as their adult counterparts (Howard 2014). Moreover, and importantly, young people vigorously supported the new law, adding their voices and agency against right-wing opposition. In the run-up to the signing of the law, Neil Howard (2014) notes that the "pioneering Bolivian Union of Child and Adolescent Workers (UNATSBO), which represents tens of thousands of under-18s all over the country, argued that regulation and labor protection are more useful for the poor and the young than wholesale bans" on their ability to work legally.

The UNCRC's idea of universal child rights is not propagated, then, without important shifts in perspective with local implementations. As Harvey (2000: 83) points out:

> It is vital to understand that liberating humanity for its own development is to open up the production of scales and of differences, even to open up a terrain for contestation within and between differences and scales, rather than to suppress them.

The notion of universal rights brings into sharp relief the tensions of scale and difference that are an anathema to the construction of a relatively homogeneous capitalist worker and consumer, which a post-war economic boom was dependent upon. Within this context, is it possible to create forms of universal rights through which "uneven geographical advancements of human interests might flourish in more interesting and productive ways" (Harvey 2000: 84)? What kinds of rules and mores may be established for young people through which we can still respect local differences and customs in a global economy where we all relate to one another in some way?

To the degree that Victorian reformers rescued a self-destructing capitalist system as they rescued children, the emergent welfare system of the 20th century faced its own crisis. Graham and Marvin (2001) outline the collapse of the post-WWII capitalist boom and the Keynesian welfare state. To the degree that Keynesian economics were tied to the modern ideal of the nation that emerged after WWII, they faced fiscal and legitimacy crises with the rise of neoliberal global economics from the 1970s onwards. As Harvey (2005) notes, global neoliberal economics' advocacy of free markets in concert with the liberalization and privatization of infrastructures (including education) delegitimized centralized state control and concomitant national welfare nets. Moreover, and importantly for geographic context of children's welfare, the narrow empirical, technical and quantitative focus of the Taylorist/Fordist regimes that dominated industrial principles throughout the 20th century has little time or space for the adult/child/community factory relations that Owen elaborated. Rather than the utopianism of Owen's factory village, Taylorist/Fordist production created an "industrial triad of the factory/village/slum of workers" within a large urban metropolitan system (Prigge 2008: 55) dominated by automobiles and machine spaces that are antithetical to the lives of children (Bunge and Bordessa 1975). With its own crisis in full swing by the early 1970s, the Taylorist/Fordist forms of production and accumulation move towards flexible (post-Fordist) accumulation, new technologies, capital reorganization at a global scale and diversification of labor relations. And, once again, young people are put at the forefront of this capitalist crisis.

In many ways, then, with its trenchant focus on the rights of the child, the UNCRC focus on individual agency, participation and flexibility played into the changing global economics of the time. Kathryne Mitchell (2003, 2006, 2017) tracks these changes in the UK, Canada and the US through education policies that move away from multi-culturalism in the 1980s and 1990s towards a focus on the globally aware student who is flexible enough to compete in the evolving neoliberal global market place. She notes that "there has been a subtle but intensifying move away from person-centered education for all, or the creation of the tolerant, 'multicultural self,' towards a more individuated, mobile and highly tracked, skills-based education, or the creation of the strategic cosmopolitanism" (Mitchell 2003: 387). These ideas are the basis of the UK's Education Reform Act (1988) and the US's 'No Child Left Behind' Education Reform Act (2002) and its follow up, 'Race to the Top' (2008), all of which emphasize standardized testing to achieve the goals of a citizenship flexibly rising to the challenges of global economic competition. Is it any coincidence that the UNCRC positions young people as capable, individuated social actors?

The debates and discussions initiated with the UNCRC in concert with the actions and activism of young people, to be articulated more fully in the second half of the book, put at stake the self-evidence of that which was and is perceived, thinkable and doable with regard to young people's places in the world. These new places of knowledge and practice break with previously understood distributions of competences, capacities, and territories. Over the last two decades, young people's rights moved from a somewhat marginalized discussion, or set of self-evident truths, in scholarship and policy-making to a central force in how we come to articulate and legislate the world. The idea that child rights are universal is sullied by discussions of migration, class, labor, and war, as well as education and development. Although not without a myriad of controversies, of which only a few are noted above, the UNCRC and its implementation around the world are largely credited for beginning the dynamic of this discussion as a global discourse with local implications. As such, the UNCRC provides a useful platform from which policy-makers and activists can argue for better children's worlds. Local shifts in perspective, then, are perhaps the best way to come to terms with universal child rights, but this is an unsatisfactory conclusion as it too stultifies the conversation. What I want to do with the next chapter is to keep the conversation going by raising an alternative to universal rights, and lay some theoretical groundwork for the empirical chapters that follow.

Notes

1 The term 'outwith' in many ways is the opposite of 'without', implying a connection – a creative tension if you will – of that which is out of our context but also in relation. It suggests a simultaneous connection and distance.
2 Owen's (1816: 81) intent was that the area outside of the school act as a pre-school for physical and moral instruction: "The area is intended for a playground for the children of the villagers, from the time that they can walk alone until they enter the school," he said, "… on his entrance into the playground [each child] is to be told in a language which he can understand … never to injure his playfellows, but in the contrary he is to contribute all his power to make them happy."
3 It may be argued that this began at the beginning of WWII with the separation of thousands of children from their parents to move them away from potential spaces of conflict. For example, in 1938, after the first pogroms against Jews in Nazi Germany, a delegation of British Jewish leaders petitioned the government to permit temporary admission of Jewish children and teenagers, with the Jewish community promising to pay guarantees. The government decided to accept unaccompanied young people up to the age of 17 years. Not long after, the government moved thousands of British children away from London and other urban areas susceptible to bombing, most going to the countryside and some sent as far as Canada and Australia (Lopatka 2007).
4 No transcript from this second hearing is available either and so it is unknown whether Jerry admitted guilt to the charge.
5 Although the 1932 Children and Young Person Act broadened the powers of juvenile courts in England and Wales, and a similar act in Scotland focused on the provision of young offenders, it was the 1969 Children and Young Persons Act that

allowed England and Wales to intervene if the proper development of the child was prevented or neglected, or if his or her health was being avoidably impaired or neglected. The same year, Scottish Social Work Act moved in a completely different direction.

6 Archard (2004) points out that in actuality the UNCRC mis-specifies what constitutes rights and what constitute children (see discussion on p. 33).

7 This cohort-specific characterization powerfully evolved at the national level to result in, to take one extreme example, people in South Africa losing rights when they turn 18 (Reynaert et al., 2015: 14).

8 I do not spend time with these articles here as they are discussed *ad nauseum* and to good effect elsewhere (cf. Kjörholt 2008; Skelton 2008).

9 On August 6, 1942, Korczak, refusing to leave the children at his orphanage, walked with them from *Dom Sierot* (My Home) to *Umschlagsplatz*, the collection point adjacent to a railway station in Warsaw where Jews were loaded onto trains destined for the gas chambers. Neither he nor the children were seen again.

3 Play, erasure, and sustainable ethics

One of the main criticisms of the UNCRC's push for universal child rights is the neglect of an alternative ethical view of the world, in which the affectionate, caring interdependence that ideally characterizes caregiver–youth relationships becomes subservient to the agency of the child and its monadic status in the world. I want to push this criticism further by advocating a standpoint that is not necessarily child-centered. A postchild perspective pushes a radically expanded conception of relationality that neither begins with the notion of monadic, individuated human beings, nor ends with the human; carers and cared for may be human and/or nonhuman, more-than-human, or an admixture of both and more. This chapter, then, is about radical relationality; it is about families and communities, the human, non-human, and the more-than-human, and the capacities afforded young people through the creation of a protected space for their becomings in and through the world.

The chapter is in the first instance about the relations between parents, children, and the state; how those are formed and erased. To the degree that the previous chapter alluded to the collusion of the family institution (and particularly the nuclear family) with the evolution of capitalist economics, this chapter focuses on families and communities as part of important vehicles for an ethical push against liberal and neo-liberal values. In short, my argument is that even though the authority and control of parents and caregivers simultaneously subverts and bolsters state neoliberal governance it also suggests a space through which something else might occur. Authority – whether parental of state – derives from complex relations, and a focus on sustainable ethics enables a form of authority that is clear, calm, bonded, regulated and weighty (cf. Deleuze and Guattari 1987: 351). I use this chapter as an opportunity to further the idea of sustainable ethics over universal rights. The last chapter suggested that rights lose their efficacy when one group's rights trump another and, as I noted, one of the main reasons that the UNCRC remains unratified in the USA is concerns over parents' rights. What I want to get to by the end of this chapter is the suggestion that sustainable ethics are a better vehicle for children's rights because they mirror a community ethics of care, and by so doing they open up possibilities for increasing what may be thought of as young people's capacities.

I am reticent to use the word capacities because of its over-use and its subsequent de-politicization in processes that push so-called capacity building from both the right and left of the political spectrum; nonetheless, from a Spinozan perspective it suggests a space that realizes and subverts limits, a place for the storage of energy and potential. I use the concept here in the sense Braidotti (2006) means when she uses capacities to speak of "Spinoza's *conatus*, or the notion of *potentia* as the affirmative aspect of power." Used in this way, it is a space that realizes the scope and potency of life. This idea of capacities dovetails with the way I am using sustainable ethics and community ethics. The idea of sustainable ethics confirms and enhances young people's *potentia*, but not in the stultifying and limited way of keeping a thing going into the future. Sustainability in this conservative sense implies a continuation of the status quo. The sustainability I am arguing for here is not about a future that we can never know but about potential in the now as an eternally unfolding moment.

The chapter looks at what sustainable ethics emanating from families and communities might entail. I argue that the family, comprising parents and/or caregivers, is an important part of democratic civil society precisely because it is not necessarily culpable to the state and it need not collude with neoliberal governance. I begin this argument below with the obvious point that the family is often where a young person first engages in relations, and extend this point to argue that relations form through the practice of erasure. To make this argument I first draw on D.W. Winnicott's (1965, 1971, 1988) ideas about play and infant development through what he called potential or transitional spaces, which I relate in important ways to Lauren Berlant's (2011, 2012, 2015) notion of living in ellipses and Braidotti's (2006, 2013) idea of capacities as *potentia*. I then make an important connection to how *potentia* is always about relations, and the importance of erasure to relations.

My previous work with Winnicott (Aitken and Herman 1997; Aitken 1998) was about dispelling constructivist ideas of development, on the one hand, and elaborating the power of play as a potently relational way of being in the world. In this chapter, I take Winnicott's notion of potential spaces and link it with how Braidotti uses Spinoza's *potentia,* as a path towards fulfillment and "gaining your heart's desire" (Harvey 2008: 23) that relates to transformation, care, and emotional citizenry. The notion of ethics of care as I use it here builds on my past ideas about the emotional work of parenting (Aitken 1998, 2009) and community contexts of care (Bosco et al. 2011), and emotional citizenry (Arpagian and Aitken 2018; Aitken and Arpagian 2018).

Living in ellipses: erasure and the creation of relational ties

Everything about sustainable ethics – and particularly as they relate to the work of parenting, creating communities of care and an emotional citizenry – is an elaboration of what relationality really means. Sustainable ethics focus on an infinite set of connections that relates everything to everything else, and the outcomes of those relational encounters (Deleuze 1983). At its simplest, when bodies or ideas encounter other bodies or ideas there is an outcome. The bodies or ideas may combine in a powerful way and become something different that is more than the sum of the parts, or one or more may be wholly or partially erased. Radical relationality accounts for a multiplicity of encounters, and things become other than what they were or how they could be imagined. At its core, and in the first instance, this form of radical relationality is a push against the idea of people as individuated, monadic, indivisible and impenetrable, which Braidotti (2013) argues comes from Enlightenment thinking and a particular ideal, as the measure of all things, which harkens back to the Italian Renaissance.

The Enlightenment's measure of all things was represented by Leonardo da Vinci's Vitruvian Man: "An ideal of bodily perfection ... [that] doubles up as a set of mental, discursive and spiritual values" (Braidotti 2013: 13). The Vitruvian ideal set proportions for architecture, technology and engineering, and it upheld a specific view of what was human, and it was not a child, or a woman, or black. This notwithstanding and as noted in the previous chapter, as Enlightenment thinkers established the notion of inalienable individual rights and radically new property and political systems that eschewed monarchies and landed entitlements in favor of equality, the formal welfare of children became important because young people could be educated into adult sensibilities, society and culture. The Vitruvian model was the perfection to which (some) children could aspire. The model set standards not only for individuals but also for societies and cultures. It created a civilization model that coincided with "a certain idea of Europe as coinciding with the universalizing powers of self-reflexive reason" (Braidotti 2013: 13). With the Enlightenment, Europe announced itself as the site of critical reason, rational individualism and universal egalitarianism. Children (and colonial possessions starting with the Americas) were viewed as the future purveyors and perfectors of this legacy.

Over the years, scrutiny of Enlightenment thinking has raised concerns about its focus on rationality, reason and paternalism, even though thinkers such as David Hume highlighted emotions over reason, passions over rationality (cf. Aitken 2009: 37–9). It is nonetheless worthwhile rehearsing concerns for the specific geometric proportionality of da Vinci's Vitruvian Man and particularly how it establishes an epitome: perfectly proportioned, encircled to reflect the celestial, and enframed to reflect the earth. He is neither woman, child, slave, machine, nor animal. Hume's focus may have elaborated the importance of emotions, and he followed Locke in thinking about the ways families are established by self-interest rather than divine intervention, but family lineage nonetheless for him is wholly male. The Vitruvian Man is "the world in miniature ... well bestowed ... composed of earth, water, air and fire, his body resembles that of the earth" (da Vinci 1883/1970). Vitruvian Man, as the epitome of human being, is the apex of development and in control of all he surveys: separated and disconnected, microcosmic and transcendent. As such, and this is the point I want to get to, Vitruvian Man's omnipotence would entail a solitude so terrifying that Braidotti (2006) argues – at least in terms of belonging and becoming in the world – this is precisely what we are not. For Braidotti, what we truly desire is not to become omnipotent but to disappear, to let life flow around us without trying to control or stop it. From Deleuze and Guattari (1987), Braidotti calls this "becoming imperceptible". This is not self-effacement, deprecation or existential alienation and anomie, but rather it is "a fundamental drive (*conatus*) ... to express the potency of life (*potentia*)" (Braidotti 2006: 7). Spinosa's *conatus* is a theory of what it is to be 'in' something. So what does it mean to be in life fully? Henri Bergson's thinking about space, time and motion helps us here.

Bergson's *Introduction to Metaphysics* (1903/1999) is in part a plea to live life fully in motion, which he calls intuition in a world of thought – emanating from Enlightenment thinking – focused on reduction and analysis. Arguing against the Newtonian and Cartesian extraction of metaphysics from science in a push towards reason and rationality, Bergson's notion of intuition presages what Grosz (2011: 1) calls "imperceptible movements, modes of becoming, forms of change, and evolutionary transformations that make up natural, cultural and political life." Bergson's point is that static positions for viewing the world are absolute and do not represent well the radical relationality, the multiplicity of encounters, in and through the material world. Enlightenment thinking resolves the subject – grasped initially through intuition – into ideas, perception, cognition, identity and a host of mental states that arrive as abstract and motionless, and perhaps no longer recognizable as parts of the whole. Every feeling and emotion, however fleeting, contains the whole past and present of the being experiencing them, and those affects are always in the moment. Bergson's point is that rather than looking at something from the outside and trying to represent it in absolute but fragmented terms, we can look at it from the inside and get in motion with it (outwith); to this extent, he elaborates a philosophy of mobility and relationality. Bergson's project to conceptualize life so that it comes to include the material universe in its undivided complexity converges with Braidotti's (2006) project to attain *potentia,* life in its fullness, through Spinosa's *conatus.* For what I want to do here, Bergson points to the ways that material and living things overcome themselves and become something different through a context of difference that is precisely about the relations between things. For young people, those relations most often start with a parent or caregiver and through intimate (and intuitive) relations that create safe spaces. To the degree that I want to get back to the family and carers, I do so because theorizing after the UNCRC very much universalized the child out of familial contexts. Moreover, I do not want to suggest a nuclear or heteronormative family can be balanced against the state as suggested by Hume and Locke, what I am suggesting here is a more radical set of familial relations that are non-heteronormative, material and fluid. In moving in this direction, I gain insight in the connections between Braidotti's posthumanism and Winnicott's (1971) ideas on play and reality, which in turn bring me to Berlant's (2015) ellipses.

Donald W. Winnicott (1965, 1971, 1988) does not separate the child from her family and environment in terms of individuation, self-discovery, objective distancing, existential angst, language, or rational detachment. Instead, he proposes a fluid, recursive process of 'separation' (that, as we shall see in a moment, is far from separation) involving intuition, experimentation and play. Rather than seeking fundamental categories of relations, Winnicott's perspective illuminates infinite possibilities by attempting to describe, at least in part, the creative processes through which children establish perspectives that reconcile, without differentiating the two (or three, or four), the seeming inner reality of the self outwith the external reality of society. Winnicott believed that recognition of a world beyond the self (and the disillusionment brought about by the realization that she may be powerless in that world) initiates for the infant a realignment of 'self' and 'object/other'. This process is not about separation but about coming into relation, and this process begins with erasure.

For Winnicott, objects are 'annihilated' or erased by the child with their removal from self-centered knowledge, but when the objects continue to exist independent of the child's awareness of them, there is the potential for the creation of new significance in relation to the child. What this means, and Winnicott (1971) was quite precise about this, is that the child creates a safe space where the favored blanket, the bunny, and/or the mother can be killed. When the blanket/bunny/mother remains intact, then the child realizes it is not omnipotent and a new relationship to the object is possible. Moreover, the object's new significance indicates to the child qualities of the self, which remain after the erasure of the object. Another way of thinking about this is that the mother acts as a container for the infant's anguish and anger, who then accepts and survives the onslaught with equanimity. For Winnicott, separation really means establishing relations anew and continuously with caregivers and with care environments. According to Winnicott, the process is the creation of a self that is different from and simultaneously in relation (rather than in opposition) to a caring other and environment. Potential spaces are spaces for the maintenance and proliferation of connections between external worlds and internal conceptions of self. They are safe places for experimentation and play. Although Winnicott's notion of potential space may seem somewhat reductionist, it nonetheless offers the possibility to mess up and make fuzzy attempts to establish clear categories of existence. Potential space, he argues, "is not inner psychic reality. It is outside the individual, but it is not the external world … Into this play area the child gathers objects or phenomena from external reality and uses these in the service of inner or personal reality" (Winnicott 1971: 51). Existing as a space that simultaneously separates and unites internal and external existence, potential space represents "a neutral area of experience which will not be challenged."

It is worth sitting with and considering the implications of Winnicott's paradox that the infant must erase their favored caregiver or object in order to enter into a relationship with them. This is true of a multiplicity of objects, people and environments, and it happens within the safety of potential spaces. The child removes objects from self-centered knowledge through agency. The survival of the object enables the child to perceive that she is not omnipotent and that the object has an existence outside of her awareness, and this is not necessarily a traumatic experience, indeed it is more-often-than-naught joyful because it leads to living life in full in the sense that Braidotti (2013) means. In this way, the child shares the world and a meaningful reality is co-created. This can only occur once the object has successfully 'survived' the child's destructive fantasies.

The child survives in the sense that Lauren Berlant (2015) means when she talks about 'living in ellipses', a metaphor about dissociation, leaps and abridgement that has an uncanny resemblance to potential spaces. Berlant defines her ellipses as spaces where the known meets what is unknowable. The problem with objects, she avers, is that they are always inadequate, which then turns back on the inadequacy of the self because the conditions of belonging cannot be presumed and always return to dissociation. For Berlant, living in ellipses is, like a potential space, not only playful but also comedic in the sense that the subject falls apart without ceasing to exist. To the degree that the child is part of the object, as noted above, the process of the object's erasure destroys too part of the child's identity. For Berlant, there is something quite comedic about losing part of the self in this kind of way. It is another, very personal, failure of omnipotence. From an elliptical (that is, indirect) perspective, the failure of omnipotence does not precipitate a crisis for the child but, rather, it pushes a comedic episode filled with laughter, giggles and frivolity. For the child, the survival of the object – whether a person, a place or a favorite teddy bear – means that it can be safely hated, repudiated, and rebelled against, but it can also be joked with, loved and accepted in a new way. Within Berlant's metaphor, events of this kind reconcile dissociation and the absurd with connectedness and the congruous. Within Winnicott's framework, events of this kind strengthen the child's self-esteem and independence out of omnipotence while fostering care and dependence.

This, I think, is where the UNCRC and universal rights lose their hold on reality. Stuart Lester and Wendy Russell (2010) write about the child's right to play through an elaboration of the UNCRC's Articles and although they raise some interesting points about play as self-protection and play as participation in everyday life, they do not sufficiently distance themselves from child-centered universalism and, importantly for what I want to say here, the child as a monadic self ostensibly removed from the family. Certainly it is difficult to reconcile the independent child and his or her agency with coherent ideas of dependency and I argue that you cannot do so without not only decentering the child but also re-embedding them in a multiplicity of relations. Moreover, the survival and continuance of these relations foments trust. Trust, in the first instance, is the confidence gained by the object's survival of the child's destructiveness. The trust foments from the safety of the potential space, and pushes towards more creativity and imagination. Unlike Freud and Lacan, who both equate desire with lack, Winnicott's perspective outlines active and positive experiences with objects, environments, desires and fantasies. Rather than a world in which there are only significant tensions between self and other – a world dominated by lacks that are debilitating and gaps that are unbridgeable – Winnicott's world is one of desire, play, frivolity, fun, connection and imagination. Importantly, from formations and assemblages that involve the child, the post-child and the more-than-child, interpretations emerge from neither the child nor the object, but from the movement of child to object and object to child. The child creates a world where desire is positive and desiring is not about trying to re-establish something that is lost forever, but about creating new and imaginative capacities. In addition, potential spaces are the safe contexts from which relations proceed or move forward. The motion is important because "movement pre-exists the thing and is the process of differentiation that distinguishes one object from another" (Grosz 2011: 1). Movement does not attach to a stable object, putting it in motion; and so movement defines the differentiation between objects. The process of movement makes and unmakes objects, including people, animals and institutions. From this idea of movement and relations comes *potentia*, and experiencing the fullness of life.

Jane Flax (1990: 116) argues that Winnicott's notion of potential spaces is an important contribution to post-Enlightenment thinking because it decenters reason and logic in favor of 'playing with' and 'making use of' as the qualities most characteristic to being human (from Braidotti and her 2013 critique of Vitruvian sensibilities, Winnicott moves us some way towards the post-human). *Potentia*, that which happens in potential spaces, highlights play and affect over logic and reason, and so it is no surprise that Winnicott theorizes that out of these arise culture and civic society (Aitken and Herman 1997). For Winnicott, potential space is a safe place for experimentation and play because it lies beyond the challenge of society's rules, and it is a place from which society's rules may be safely challenged. The ideas of potential and transition describe a space through which both the child and society are treated as dialectically interrelated moments of action. Culture, politics and symbolism are not immutable structures that necessarily define children, but rather children contest these structures in the process of creating cultures, politics and symbolisms (Aitken and Herman 1997). Put another way, culture is not seen as Freud's external and coercive 'law of the father', which forces the child to separate from the mother and embrace an abstract and immutable patriarchal culture through complicity with what Lacan (1978) calls the *big Other*. Rather, there is the possibility of the child bringing something of her inner self to the traditions and practices of a culture in order to be able to make use of them. In this account, the agency of the child shapes her cultural practice and vice versa. Also, according to this account, the infant's ability to choose and utilize transitional objects begins the process of symbolization.

For Winnicott (1971: 102), the capacity to play and the process of symbolization expand "into creative living and into the whole cultural life of man [sic.]." Culture and politics, like play, are not only something that the child can 'make use of', but they are also traditions to which she can bring her inner self. Does this open up the possibility of saying something about relations between children, families and the state? Is there something about Winnicott's notion of erasure and Braidotti's notion of *potentia* that come together in ways that help elaborate the machinations of the state apparatus, which overtly or inadvertently define the citizen-self?

There are a number of interesting theories on the politics of play, which suggest creative and destructive tensions between individuals, cultures and the state (cf. Huizinga 1955; Geertz 1973; Schechner 1993; Sutton-Smith 1997; Henricks 2015). From a synthesis of these theories, Tara Woodyer (2012) usefully articulates play not only as an everyday practice but also as a political practice that exceeds representation. Drawing on research by Cindi Katz (2004) and some of my earlier work (Aitken 2001), she argues that play is not just about social reproduction or rehearsing social roles, it is also about power and control, about coming to consciousness and becoming other: "Playing works through aspects ... that are somewhat mysterious; identities, social relations, and socio-material practices are played with as details are tweaked or wildly (re)imagined" (Woodyer 2012: 318). Importantly for what I want to say about youth activism later in this book, Woodyer goes on to note that when players engage in so-called acts of resistance they often express themselves through enjoyable activities. Moreover, from a posthumanist perspective, she argues that play prioritizes "the non-cognitive and more-than-rational" because of "its embodied nature, its heightening of the affective register, its momentary temporality, its intersection between being and becoming, and its intensity" (Woodyer 2012: 319). This context-dependent process of becoming is also central to Winnicott's notions of play, but in what sense is this transformative potential linked to the power of families and the state? Flax (1993: 122) pondered a similar question, wondering how ideas of culture and politics play out in state-mediated, public and political contexts: to the degree that "justice must be held and nurtured within public, political-economic contexts rather than the dyadic ones of early childhood" is it possible to create a play of justice, law and rights? Flax argues that our capacities to seek justice arise in part out of potential spaces and the lifelong processes that Winnicott describes. The design and engagement of just practices and ethical living offer the possibility of new modes of relating with others and alternative ethical practices. If justice and rights are better understood as ongoing processes rather than as fixed procedures or pre-given universal standards to which all must conform, then they also help us reconcile and tolerate difference without recourse to domination. It seems worthwhile to consider in a bit more detail, then, how difference shows up as an ethical focus for belonging and becoming, and how it relates to individuals, families, communities and the state

Difference-centered ethics

Before focusing on family authority as a push against the excesses of state capitalism, it is worth considering how some larger tensions play out between children, families, communities, and state relations. Bosco and his colleagues (2011: 159–61) identify two series of tensions that revolve around civic relations. The first revolves around individuated versus communitarian concepts of civic relations. Neo-liberals focus on individuation, building on the contractual ideas of John Locke (1690/1824, 1693), discussed in the previous chapter, and John Rawls' (1971) ideas of distributive justice to establish citizenship largely in terms of individual ownership of property, legal rights and privileges. As noted earlier, when first conceived by Locke, this notion of citizenship disenfranchises most women and all children. It positions young people as not yet competent to enact civil responsibilities and relies on a paternalistic society to help teach and educate them into competency and autonomy. In time, all citizens who own property through their individual labors are equally competent for civic participation, and gain requisite legal rights and privileges. Children can learn how to labor and derive the capacity to be autonomous. Communitarian perspectives draw on the 19th nineteenth century writings of Catherine Barmby (cf. Sanders 2001) and Robert Owen (1816/1972), and from Anthony Giddens' *The Third Way* (1998). This perspective lauds values and beliefs that exist in the public sphere, and adheres to the notion of group membership rights. Women and men co-exist as part of the collective, but children are seen as dependent, not-yet-citizens embedded within a social group. From this perspective, membership within society is also realized through participation in civil responsibilities.

A second series of tensions revolves around expectancies (Bosco et al. 2011: 160). Within the individuated perspective and the communitarian perspective the expectation is that men and women are responsible and autonomous in their decision-making. In both perspectives children are dependent, attaining autonomy only with maturation. A third perspective on citizenship, cosmopolitanism, focuses on the creation of globally competitive citizens. Arising from Immanuel Kant's notion of a common humanity it finds a contemporary champion with Martha Nussbaum (1997). The cosmopolitan citizen is expected to be nationally patriotic and globally aware. Moreover, to the degree that Nussbaum's (2001: 11) ideas feed on the tenets of universalism, the expectation is that "universals ... are facilitative rather than tyrannical, that [they] create spaces for choice rather than dragooning people into a desired total mode of functioning." Nussbaum's ideas of universalism raise the self as an individual responsible for her or his own decisions, which plays into contemporary ideas of neoliberalism, and her cosmopolitanism rests on the shoulders of a globalized world.

All three of these perspectives on civic participation – individuated, communitarian and cosmopolitan – embrace aspects of relational identity formation but it is quite clear that all three fail to treat children as other than becoming adult citizens. They cannot accommodate children as people who collectively participate, who are already competent at a number of civic duties and who already fulfill most expectations for citizenship. Flax (2001: 26) takes an ethical stance against Nussbaum's cosmopolitanism that is also pertinent for individuated and communitarian perspectives; she is "unpersuaded that there is any such good for all humans, or that this search for the good, however conceived, is the definitional or even a particularly accurate way to characterize human experience." By taking this position, Flax is not arguing that ethical inquiry is impossible or unimportant, but that there are other ways to practice it (cf. Flax 1993).

One possible alternative for understanding children's (and adult's) place in the world, and their rights, is through the lens of affectionate, caring interdependences that characterize ideal caregiver–youth relationships. Mehmoona Moosa-Mitha (2005) defines the citizen–self as a relational, dialogical self, who gains a sense of self through relationships and dependencies on 'others,' including people, places and events. This is not, she points out, the so-called responsible self of neo-liberal theorizing that is found in the contemporary rekindling of individuated and communitarian perspectives. Her citizen-self is difference-centered and responsive but not necessarily responsible for his or her context along known lines of normative expectations writ large in terms of civic duties and obligations. Equality in these terms is predicated upon sameness. From the Rawlsian neo-liberal perspective, those who are not competent in the use of reason, such as young children, cannot lay claims to being equal. Alternatively, a difference-centered relational perspective posits belonging as fundamental to interpretations of equality, which is guided by subjective experiences rather than objective notions of rationality and reason. Elaborating Yuval-Davies' (1999) notion of "differently equal," where equality is defined through difference rather than sameness, Moosa-Mitha (2005) argues that freedom is the right to participate differently. In this sense, rights are defined broadly in terms of their multiple relations and dialogues with society. And so, children's participation and citizenship, like everybody else's, must be examined in terms of interdependent relationships rather than in terms of universal rights, or rights of autonomy that emphasize independence from these relationships. Moosa-Mitha redefines children's rights of freedom, in a relational way, by examining whether children are able to have a presence in the many relationships within which they participate. By presence, she means the degree to which the "voice, contribution and agency of the child is acknowledged" (2005: 381). Presence, more than autonomy and individuation, acknowledges the self as relational and dialogical. It is not enough that children have a voice; they must also be heard in order for them to have a presence. Not to recognize the presence of a citizen is itself a form of oppression. To get beyond the impasses in addressing children's rights and responsibilities, then, Moosa-Mitha's difference-centered approach is summarized by two particular axes of recognition. The first introduces the notion of the citizen as having active selves with not only agency but also capacity. The second defines the citizen-self as relational and dialogical, who gains a sense of self through relationships and dependencies on others, including people, places, and events. This is not, she goes on to point out, the so-called responsible self of neo-liberal theorizing that is found in the contemporary rekindling of individuated and communitarian perspectives, but rather it is a relational self, connected and contextualized. The difference-centered self is responsive but not necessarily responsible for his or her context along known lines of normative expectations writ large in terms of civic duties and obligations. A third axis may be added, which focuses on differentiations and relations that are spaced and scaled (Aitken 2014).

In later work, Moosa-Mitha (2017) recognizes this third axis by tackling the ways that children's citizenship is positioned through the UNCRC's assertion that it works in the 'best interests of the child.' She argues that those interests are mitigated only in specific geographic spaces and at particular scales. Children's citizenship, she argues, is most often and problematically contextualized socially, which provides a narrow understanding of relationalities that rarely consider the spatial or political. Children's best interests, then, are defined in terms of social welfare most often contextualized through the home and community. Dominant discourses of children's best interests, according to Moosa-Mitha (2017: 3), then, "assume a universal and fixed age-related identity of childhood that overlooks the diversity of identities that children have and the difference that this makes to their experience of belonging" in particular places.

In terms of current neo-liberal and neo-welfare agendas, as Moosa-Mitha (2017) notes, the rights of children to participate freely in society are usually ascertainable only within specific socio-historic and geographic contexts. Although children's rights to cultural- and self-expression are enshrined in the UNCRC, more often than naught at the local level children are constructed as less-than, not rational or mature enough, and their participation is overlooked or legislated against. Children are dependent, but to cast them as solely this, or solely needy or irrational, or a combination of those things, casts them as incapable of participating and is a result of certain liberal theories that posit the citizen's self as independent and autonomous.

Returning to the radical critiques of the UNCRC discussed at the end of Chapter 2, it is clear that beyond their identities as children, other axes of difference such as gender, race, identity and sexual orientation will lead to differential experiences. To the degree that UNCRC's focus on provision, protection and participation is thinly veiled in white, adultist, heterosexual norms and practices, then it is important to adopt difference-centered relationalities. These formulations garner inspiration from Iris Marion Young's (1990) celebrated politics of difference, which takes them a few steps further by addressing complex issues not just about distribution but about "decision-making power, division of labor, being disadvantaged by dominant norms, and freedom of cultural expression" (Young 2006: 91). A difference-centered approach that raises these issues in the context of children's rights and freedoms addresses presence primarily in terms of young people's capacities as participatory beings, as well as examining the normative assumptions of social institutional practices that bar that participation by not listening, by overlooking, by tokenism, or by actively condemning. Together, the work of Moosa-Mitha (2005, 2017) and Young (1990, 2006) helps us move towards a notion of rights that is fluid and relational. Difference-centered theories of rights help position children's multiple and fluid relations within family, community and other institutions. If we are interested in difference-centered theories as they relate to families and state relations then it is important to understand the contexts of authority and how that plays out in those relations.

Familial authority and state relations

In Chapter 2 it is noted that Locke's attack on the divine pre-determination of authority is also an attack on the patriarchal authority of the family and the state. What bears looking at in this chapter is David Foster's (1994) argument that Locke (1690) not only rejected patriarchal rule at the level of the state but also father's rule in the family. Locke (1693: 106–8) argued parental authority should focus on rewards and friendship but also shaming rather than beatings and harsh language. His focus on familial desire revolves around men desiring the act of procreation (Locke 1690: 34), a statement rightly criticized by feminists pointing to Locke's continued ties to the masculinism (if not patriarchy) of Enlightenment thinking. Feminists also argue that Locke's continued differentiation between the 'private' family and the 'public' state creates a spatial frame, an artificial geography and a problematic binary, which forecloses upon the gender politics of families and leaves patriarchy intact (cf. Eisenstein 1981). Countering this, Foster contends that Locke's intent was simply to argue that the relations between children and parents is a private matter, and fathers should not act like kings. In addition, in the public realm, kings should not act like fathers. As I argue elsewhere, Locke's point is not that parental power and state power constitute private and public realms respectively, but rather that their separation presupposes a different understanding of each (Aitken 2009: 35).

Locke heavily influenced David Hume's ideas on authority and the family but Hume took these ideas further in his famous elaboration of the ways passions trumped reason. In his pursuit of radical skepticism, while understanding the importance of rational insight and empirical validation, Hume (1739/1955: 415) nonetheless lauds emotions over reason, famously saying that "[r]eason is, and ought only to be the slave of the passions, and can never pretend to any other office than to serve and obey them ... A passion is an original existence, or, if you will, modification of existence, and contains not any representative quality". To the degree that emotions are important, how then do they show up in the relations between children, parents and state, in terms of contexts of authority? Hume's point is that although reason is important, the role it plays in relations is purely and solely instrumental in that it teaches how to establish authority, but the authority itself is defined through emotions. The overriding force that guides action is not reason, a sense of obligation, or an innate morality, but the desire for what he calls self-gratification (Herman 2001: 200). Self-gratification, for Hume, resided within a society and a form of governance that propagated good habits. Herein lies Hume's understanding of the relations between self, family and the state.

The presumption that authority of parents is different from the authority of the state and its fundamental basis is emotional rather than rational leads us to a consideration of relations, justice and the law. This presupposition requires a different way of thinking about how children, families and the state relate, and for what I want to do here that way of thinking is best rooted in positive desire and *potentia*. The problem with Locke's Enlightenment thinking is that his primary desire is self-preservation, which as we noted in the last chapter is parleyed into property rights, and his secondary desire comes through propagation, which is parleyed into the right to family life. As critical as these designations were for dismissing patriarchal and divine rights, and as I have argued elsewhere (Aitken 2009: 36–8), they nonetheless sidestep the messy relations between families, children, and the state by prescribing mechanistic and multi-scalar forms of justice and universal rights. Hume moves us in the right direction by suggesting that while desire is the most important part of the establishment of authority, it is realized through emotions and the primary desire of self-gratification. There is no moral authority behind this gratification but it is tempered by society and government through what Hume called good-habits; that is, it is ethically based. It seems to me that this leads us down a road that ends with how I am conceiving sustainable ethics. But we are not quite there yet.

From Hume, the classical way of thinking about the state authority and law is that it should be fair, just and equal, and its authority should apply to everybody. Nobody is above the law, but parents and caregivers are responsible for their own children far more than children in general, and so there is a differentiation of authority and as many styles of care as there are caregivers. When those relations are broken, and the state subverts the authority of parents (as, at times, or with particularly abusive caregivers, it should) it is not just a question of parents' rights or children's rights, it is a question about the reordering of civil society. The irony of the *In re Gault* (1967) case, from the previous chapter, for what I want to talk about here is not that Jerry Gault was denied the rights of *habeus corpus* afforded adults or that he was remanded to a State Industrial School for six years when an adult on the same charge would have received a maximum sentence of two months in jail and a $50 fine, but that he was arrested without the knowledge of his parents. That Jerry Gault's parents were not notified of his arrest and did not hear of it until they were contacted by his co-defendant's parents may simply point to a lapse of judgement or a lack of care on the part of the police but to the degree that it set in motion a court case that reified the idea of children as separate from adults and the subjects of their own sets of rights, it is important. That Jerry

Gault's parents were not contacted immediately (it seems unreasonable to assume that police could not have found one of them at work) speaks volumes about state disregard for familial relations. The issue here is that a focus on children and their best interests prescribes an individuated, one-sided context for justice and the law. It speaks also to the rights of parents in relation to the protection and provision of their children, which is often stated as a reason why the USA has not ratified the UNCRC. The UNCRC codifies that children have the right to protection and provision, and the implication is that if children are not provided for and protected within a family unit, then the state will intervene. Some argue that from WWII onwards and the evolution of the welfare state there was a swing towards the protection and provision of children first and foremost by the state rather than the family. In terms of necessity, nonetheless, the remits of Child Protective Service in the USA and the National Society for the Prevention of Cruelty to Children in the UK are complex, yet both elaborate in their mission statements a wish to maintain the integrity of a child's home and family if at all possible. That said, the argument from state institutions such as these is always for the best interests of the child and that "every childhood is worth fighting for" (NSPCC 2018). That the influence of the Western family as an institution declined after WWII is well documented, and debates raged in the 1990s about what that might mean (cf. Aitken 1998). I am particularly persuaded by the arguments of Jean Bethke Elshtain (1990) who points out that the primary problem with the decline of familial authority is precisely because it is special, limited, and particular. The particularity of parental authority means that it may be abused in ways more insidious than any other kind of authority, but without it parenting would not exist. Family authority is imperative, for Elshtain, within a democratic, pluralistic order precisely because it is not necessarily homologous with the principles of civil society.

Work from the new field of ethno-psychology comes to the same conclusion but from a different perspective. From the standpoint of the "new science of child development" that draws from evolutionary biology, Alison Gopnick (2016: 22) notes that "[p]arents are not designed to shape their children's lives. Instead, parents and other caregivers are designed to provide the next generation with a protected space in which they can produce new ways of thinking and acting that, for better or worse, are entirely unlike anything that we would have anticipated beforehand." Clearly there are important relations afforded by the maintenance of familial and caregiver integrity. The idea of increasing young people's capacities through sustainable ethics – their *potentia* through *conatus* – is clearly best credited to the provision of protected spaces (locatable but not necessarily prescribed) for development, surprise, growth, dislocation, nourishment and becoming other. When human/parent/child rights get in the way of that (that is, when they are framed so as to foreclose upon becoming other) then capacities are diminished.

If potential spaces and the *potentia* that arise from them are theorized at the level of the state and culture (and Winnicott certainly advocated the latter), then we need to think more carefully about how that plays out for ideas of justice, rights and the law. Flax (1993: 122) argues that Winnicott's ideas are an antidote for the instrumentality of legal cultures: "Both our capacities to seek justice and our needs for it arise in part out of transitional spaces and the lifelong process Winnicott describes." Rights, justice and the law, she goes on to note, are held and nurtured within public, political-economic contexts rather than in the dialectical processes of early childhood. The potential for processes of justice (including rights practices and the enactment of laws) "as intermediate areas of experience in which relief is found from the strain and isolation of our inner worlds and the press of necessity and regulation from our outer worlds" suggests a way out of the seeming instrumentality of many legal systems. This may sound overly reductionist, but it also suggests that engaging in just practices (and good ethical habits) fomented from sustainable ethics rather than rigid or universal rights offers the possibility of enriched relatedness with others. Justice and rights become fluid and playful rather than a pre-given standard to which everybody must conform. Children are neither objects of rights nor subjects of necessity; in play there is no subject or object. Processes of justice and sustainable ethics as ongoing practices can turn necessity (caring, bridging gaps, protection, participation) into a source of pleasure, creativity and play. These processes can establish a space of safety and security while expanding capacities. Fear of losing what I have or not getting what I want collapses my world into something rigid and monadic. Flax's (1993: 123) notion of justice teaches how to reconcile with and tolerate difference without domination: "It generates processes in which we can differ with the other without feeling the need to annihilate her."

Potential spaces enable people to engage safely in a continuous struggle to accept first the mother/father/bunny and later, others, as existing separate and yet in relation to us. In Winnicott's best possible world the mother and child are invested in the best interests of the child's development and becoming. The mother and the child were once biologically connected and through potential spaces each gets to appreciate the separateness and imperfections of the other. A question that is worth posing as I try to raise this discussion to the level of the state, is what happens when the mother tries to erase the child. Winnicott (1965) posed the idea of the good-enough mother/environment. The good-enough mother/environment makes it possible for the "individual to cope with the immense shock of loss of omnipotence" (Winnicott 1965: 95). The good-enough mother adapts to an infant's needs at first and then gradually adapts less and less as the infant is able to deal with her lack of perfection (Winnicott 1953: 92). The good-enough mother gives the infant a gradual sense of separation rather than the sense of being abandoned or dropped or, worse, projecting the shock of annihilation. The good-enough mother, obviously, does not annihilate her infant, but there are mothers (and fathers and other carers) who cannot cope with their young children, who emotionally abuse them, who give them away or worse. This kind of erasure happens at the state level also. What, then, comprises the good-enough state?

Living in eclipses: erased by the state

Like the good-enough mother, the good-enough state affords a safe space through which citizens can simultaneously play with relations, accommodate difference, exercise creativity, and engage in just practices. For Flax (1993: 127), the rule of law alone – no matter that it is grounded in ideas of natural rights, reason, moral imperatives, or communicative competence – can never create a good-enough state. Justice, just practices and good habits foment from sustainable ethics. From Hume, it is reasonable to assume that these ethics are based on emotions or sentiment rather than abstract moral principles. The kind of justice that evolves from the authority of good-enough parents and good-enough governments is a potential space that help us play with, tolerate, accommodate, appreciate, and remake. Justice, then, is a space through which we can not only come into a variety of relations anew, but through which we can also remake authority and rules. It also encourages us to renew and remake ourselves, and how we show up in the world. Justice is necessarily an outgrowth from active notions of citizenship (Flax 1993: 125). This flexibility is lost when justice becomes rigid, and the form of just practices is mandated from the top down. To live in a state of fluidity, renegotiation, and collective agreement is to live in a state of safety under justice, law and sustainable ethics. The power of laws or rules – or put less severely but no less poignantly, Hume's good habits – lies in a collective agreement to follow them as best we can. This is the safe and protected place from which *potentia* flourishes. For this to work as a celebration of difference, the state must agree to include other persons within these kinds of protected spaces. If this kind of agreement is not in place, then it is an implicit rejection of the processes and practices that such agreements codify and, as Flax (1993: 127) presupposes, most likely "some of us will not be secure against the aggression of others." As Agamben (1995, 2005) pointedly notes, the modern state since WWII has shown time and time again its capability of defining some groups as bare life, or less than fully human, and hence beyond relations of justice or the protection of the law. At its worst, then, this is a less-than-good mother/state attempting the annihilation of erasure of the child/other. If, for Berlant (2015), to live eliptically is to live the fullness of life in the present as proximate and connected but not necessarily embedded or immersed, then erasure by the state is to live in eclipse. Her eclipse metaphor is about darkening and obscurity, where a justice and legal status are removed for particular peoples.

Why, precisely, are certain people targeted for the dispossession or eclipse of rights? One of the ironies of the development of human rights since WWII is that there seems always to be people who are not included or whose rights are trumped by others' rights. People perceived as disorderly, fearsome or other because they do not fit prevailing societal sensibilities are often shunned or ostracized from national culture and politics, and marginalized to peripheral and inadequate spaces. Pulling from Roman law, Agamben (1993, 1995) famously theorizes these people as akin to the ancient *homo sacer*, or sacred man, who may be killed with impunity as long as the process was not sacrificial. The *homo sacer* was less than human, and even ritual sacrifice was too good for him as it might offend the gods. Denied any kind of political status (*bios politikos*), the *homo sacer* was left with bare life (*zoē politikos*). With the rise of democracy emanating from the Greek *polis* and Roman *agora*, Agamben notes that the idea of the *homo sacer* increasingly made little sense, but with modernity and events like the holocaust, hunger strikes, Ghandi's privations and so forth, bare life is back and biopolitics are now a foundational part of the state apparatus. And, with the modern state, come spaces – from Nazi concentration camps to Abu Ghraib and Guantanamo Bay – circumscribed by bare life. As noted in Chapter 1, Agamben elaborates the function of these spaces and bare life in modern politics through "the state of exception" (1995: 9) whereby citizens are protected by the sovereign rule of law and certain outsiders (the modern *homo sacer*) are purposively excluded.

In addition, drawing on the work of Agamben and an expanded view of his states-of-exception through Aihwa Ong's (2006) work on contemporary neoliberal governance, I argue that young ethnic minorities are particularly vulnerable to the erasures that are hidden under state policies of seeming democracy and justice. And, drawing on recent work on youth geographies and citizenship, which I contextualize through Étienne Balibar's (2012) "impossible communities of citizens" and James Holston's (2009) ideas around "spaces of insurgent citizenship," I argue further that it is from young people's actions – and especially their resilience and revolutionary play in dystopian urban spaces – that hope for communities of care and sustainable ethics arise. For Braidotti (2013: 93–4), this kind of sustainability is found only in a grounded, situated and a very specific feminist politics and the ethics derive from what she calls – drawing from Agamben – zoē-centered egalitarianism. For Braidotti, as explained in the first chapter of the book, zoē is not about some basic animality, it is about the relations between children, adults, women, men, technology, institutions, animals, microbes, stones and mortar that make us who we are, and more-than-human. The rise of postchild sensibilities described in the first chapter of this book – in conjunction with posthuman thinking in general – recognizes young people's relations, ambiguities, dependencies, autonomies, and politics. It understands that at any one time and at any specific place the actions, practices and politics of young people are an assemblage of relations with other young people, technologies, adults, animals and other materialities that cast doubt on the nature of being and becoming. In the next two chapters, I examine the contexts of erased youth denied citizenship rights on the basis of ethnic identity, and how that denial contextualizes itself for them with their families, communities, and in modern and independent Slovenia.

Towards sustainable ethics

To the degree that family relations and the authority of caregivers (including non-human caregivers) are imperative within a democratic, pluralistic society, they are also never absolutely homologous to that society. Indeed, a direct and hierarchical relationship between the private lives of children and the public polity would perforce weaken democratic principles. The examples of erased Slovenian families that we are about to encounter point to some of the implications of a public polity that inadvertently pits children against care-givers. Elshtain (1990) asserts that children need particular, intense relations with adults. Such relations enable children and families to oppose and contest societal norms as a marker of what works in democracy. Elshtain maintains that respect for difference between parents and children in families and between families and society is crucial for the maintenance of democracy: "The social form best suited to provide children with a trusting, determinate sense of pace and ultimately 'self' is the family ... it is only through identification with concrete others that the child can later identify with nonfamilial human beings and come to see herself as a member of a wider human community" (1990: 60). The discussion of family and state relations that I enlarge in the next two chapters using the context of Slovenia's erasure of ethnic minorities as a strategy targeting certain groups for de-politicization, highlights how survival pushes *potentia*, and the importance of trust within ethical communities for families dispossessed of power.

Concepts like rights, due process, and equality are grounded at least in part in certain ideas about innate or essential human properties. Leonardo da Vinci and his Vitruvian man informed 18th century Enlightenment from a 15th century context of essentialism, and it was that inspiration that propelled large swathes of the industrial and scientific revolutions. But there is much more to it than this and as such I am hesitant to fully map contexts of indi-vidualism, monadism and the perfect (hu)man that come from this time onto Enlightenment thinking. One of the problems, as I noted in the previous chapter, is that Enlightenment is so closely tied to the scientific revolution, and its inherent rationality. From the Enlightenment thinkers of the late 17th century onwards, and especially Locke, Hume and others who thought about children, families and the state, there is a sense that perhaps the instrumen-tality, monadism, rationality, and reason were not as central to their thinking as later scholars suggested. As Deleuze (2001: 35) notes "David Hume pushes the furthest ... His empiricism is a sort of science fiction universe *avant la lettre*. As in science fiction, one has the impression of a fictive, foreign world, seen by other creatures, but also presentiment that this world is already ours, and those creatures, ourselves." This science fiction is supposed to compel the state to act as part of a contractual obligation to protect a set of rights that it did not create. There are, of course, many reasons to lay the problems and limitations of essential human rights at the feet of the Vitruvian Man and Enlightenment thinking. In noting this Flax (1990: 230) suggests that no reasonable alternative seems to exist. This, I think, is a problem. What I offer as we move forward to discuss some of the excesses of state culpability in the next two chapters is an alternative based upon Moosa-Mitha's (2005, 2017) difference-centered relationality and Braidotti's (2013) notion of sustainable ethics. Harvey (2000) argues for the possibility of universal rights out of two foundational contexts: the production of spatial scale and the production of geographical difference. Out of this, perhaps it is reasonable to argue, as a radical alternative, for a shift from universal child rights that emanate from the Global North and Western societies in the 20th century to a push for sustainable ethics from young people in the Global South and Eastern Europe. This transformation moves child rights from adultist proclamations thinly veiled in Western/Global North normativity to a form of local dissent and disruption out of communities of care, and primarily led by young people.

4 Codifying state erasure in post-independence Slovenia

Much has been written about what happens when a state singles out a particular group for legal annihilation, and yet as often as not each situation is unique and to some degree incomparable with others (Bosco 2006; Yezer 2011; Zucker 2013). The Slovenian erasure was not an extermination, some point out, while others argue that it was never planned and was primarily attributable to regrettable legal and bureaucratic fumbles. Against these arguments others point to a systematic and subtle ethnic cleansing, and to the degree that the state was culpable, it must be recognized that it took over 10 years for some kind of rectification to begin, and this did not happen until there was intervention from the European Union's Court of Human Rights in Strasbourg. Human and social rights have an important place in democratic societies to the extent that they are a legal basis for pointing out inconsistencies and abuses, a way to converge on some level of shared accord, and a hook to which activists and legislators can attach arguments, but my enduring concern is that they do not translate seamlessly to local situations. For that, I argue, situational and sustainable ethics are required. I set up theoretical delineations of sustainable ethics in the first chapter of the book, which I picked up again in the previous chapter, but before I can get to a fuller elaboration of what a locatable ethic of sustainability means practically some preliminary questions about state policy and local implications must be asked. How does a specific state get to set in motion a process of annihilation? What are its ramifications for the targeted people? How do those ramifications play out for different segments of the erased population? How do they show up in the family and amongst young people? What was it like to spend your formative years as an erased person? In what ways can states be held to account? I try to answer these questions in a first go around with a discussion of the erasure of ethnic minorities after Slovenia got its independence, but what I offer is at best a mere glimpse of what young Slovenians and their families went through.

The effects of the changes in citizenship and residence rights on certain young people with the enactment of the Slovenian Constitution and, specifically, its *Alien Act* in 1992 is an important example of erasure from a number of interrelated standpoints that go beyond one country, an isolated event, and a singular context of alienation. The Slovenian example speaks to larger concerns about unstable citizenship regimes and the precariousness of youth rights under the tutelage of emerging neo-liberal economies and modes of governance. It suggests disturbing parallels to contemporary immigration problems plaguing the USA and Europe in particular but also in East Asia, Australasia, as well as internal migrations in China and India. Further, the Slovenian problem is about the creation of a country, of its relations to nationhood and nationalism, and how civic participation and youth rights of citizenship intersect with local and global transnationalisms. I think this is important for understanding how to locate an ethics of sustainability. The Slovenian erasure case, its progress through the courts, and its on-going solution speaks to the creation of constitutional law and also the larger, global issues of the evolution of human rights discussed in Chapter 2, and their concomitant and problematic relations with child rights and young people's activism. An enlarged discussion of youth activism is provided in the next chapter. I use this chapter and the next to not only introduce a poignant example of state erasure but to underline the hugely problematic tensions and diminished capacities that arose within families as a result of this draconian policy. This chapter details the context of Slovenian independence and the creation of constitutional events that elaborated special circumstances for ethnic minorities living in the country at independence, and especially children who were born there. The empirical arguments are drawn from interviews with children and adolescents, family members, lawyers, activists, journalists, filmmakers, and academics living in Slovenia, initiated in 2014 and continuing as I write this book.[1] I use what effectively was a de-legitimization of particular groups of people (some refer to it as political genocide), touted by some as the worst human rights abuses in contemporary Europe, to position a practical example that bolsters my main argument for the importance of the right for young people to make and remake space, and also themselves. These arguments are carried, albeit in a different way, with examples from elsewhere in the world in Chapter 6.

The Slovenian state

The complexities of the Slovenian problem involve migration, language, belonging, identity, alienation and exclusion. I will attempt to elaborate and disentangle some of these complexities in this chapter and the next. The story of the Slovenian erasure is, quintessentially, a story of evolving self-determination for a particular people and, in short, it goes as follows. As the most northern of the Balkan states, Slovenia began gaining a sense of self in the 16th century when its mostly peasant language was raised as something that defined its people as special. By the 20th century, a cadre of poets and writers had secured the defining place of the Slovene language in the arts, humanities and, to a degree, the judicial system. WWI intervened to stop any kind of nation-building; perceived as the epicenter of the war, the so-called South Slavic Peoples was created out of three provisional nation-states comprising Slovenes

Croats and Serbs, and renamed the Kingdom of Yugoslavia in 1929. In this sense, unification was seen as a way to counter some of the regional tensions that began WWI's conflict; the opposite of 'Balkanization,' as the division of a region into smaller and mutually hostile units. After WWII the Kingdom of Yugoslavia – with a territory that encompassed Slovenia, Croatia, Bosnia, Serbia, and Herzegovina, Macedonia and Montenegro – was renamed the Federal People's Republic of Yugoslavia. Born to a Slovene mother and Croat Father, Josip Broz Tito became prime minister in 1944 and president in 1953 until his death in 1980. Tito's policies strove to unify the ethnic divisions within Yugoslavia and undermine Balkanization even more under a *bone fide* national *raison d'être*, and to a large degree he was successful. With no borders between the Federal republics of Yugoslavia, there was free flow of goods and workers. An expanding manufacturing base from 1960 to 1980 required more input than Slovenia's 1.6 million people (1960) could supply. Under Tito, the Yugoslav people were relatively mobile, and Slovenia was attractive for southern workers interested in well-paying jobs. It was the wealthiest republic with growing industries and commercial enterprises. The post-WWII decades witnessed economic growth and a steady migration of people from the other Balkan republics, some of whom intended a complete change while others maintained ties to their homes to the south. Slovenia recognized these long-term workers who established families under law as *animus manendi*; that is, it was understood that they intended to stay (Zupančič 2016). Their children were born into the Slovenian language, culture and economic system. Often, these young Yugoslav citizens had little knowledge of their republican citizenship because prior to 1991 connection to a republic had no legal consequences (Zorn 2011: 70). Tito died in 1980 and the seemingly harmonious Yugoslavian character began to unravel. By 1990, it was clear that the more populous Serbian people (or, more correctly, populist politicians such as Slobodan Mulošović) harbored a nationalized agenda to unite Yugoslavia under Serbian control.

Slovenian politicians were unhappy with what seemed like a Serbian push to control Yugoslavia. A plebiscite of all people living in the Republic of Slovenia in December 1990 – which included a significant number of permanent residents from other republics – overwhelmingly (86% in favor with a 90% turnout) supported independence. It is unlikely that the ethnic minorities voting for independence were aware of what would result from the legislative machinations of the new state. After a short-lived 10-day war with the (now mostly Serb controlled) Yugoslav army, the Republic of Slovenia separated from the federal state. The focus of the Yugoslav army went south to Croatia, which had also recently declared independence but had more Serbian enclaves. Thus began the 1990s Balkans' war, out of which Slovenia got off very lightly compared to it southern neighbors. The status of the people from other republics who had migrated to Slovenia over the last 20–30 years came into question. Ethnically they were Serbians, Croatians, Bosnians, Roma, and some Albanians. Many did not speak Slovenian (although their children did) and the local population often perceived them as foreigners. Under the strictures of the socialist bureaucracy, they were well enumerated and listed in official documents and registers of residence, employment, taxes, social security and so forth. Most had been issued Slovenian identity cards and passports in addition to their Yugoslav documentation. Given this enumeration, it is curious that when a six month window, from June to December 1991, was opened for those born in other republics and their children to apply for Slovenian citizenship very little notification was given (Mekina 2014).[2] Within this time period 170,000 applied for and got citizenship, and on February 26, 1992, all citizens of former Yugoslav republics who had not applied for Slovene citizenship lost all their legally acquired status of permanent resident; they became known as the *Izbrisani*, the Erased.

In what follows I outline the legal building blocks that led to the seeming arbitrariness of the Slovenian erasure before raising specific stories of isolation and incomparability that challenge the architecture, legality, and length of the erasure.

The architecture of the Slovenian erasure

To appreciate the differences and distinctions in erasure stories I discuss in a moment, it helps to begin with an understanding of some of the history of Yugoslavia that resulted in so many ethnic minorities moving to Slovenia. It also helps to understand the architecture of the erasure, which led to capricious decisions by low-level bureaucrats.

On May 15, 1991, a constitution was drafted that included the *Citizenship of the Republic of Slovenia Act*. Article 40 of this act defined the conditions under which immigrants from other Yugoslav republics could obtain citizenship, which included having registered permanent residence in Slovenia on the day of the plebiscite. An application with the closest administrative unit within six months of the Act's enforcement was required, but this information was not widely distributed, nor was the fact that failure to apply meant that not only was Slovenian citizenship denied, but legal residential status was also lost (Mekina 2014). The Act became law with the declaration of independence on June 25, 1991, and by December 25, 1991, 172,000 people from other republics (8.6% of the population), of whom around 30% had been born in Slovenia, applied for citizenship (Šalamon 2016).

In addition to the *Citizenship of the Republic of Slovenian Act*, another fundamental constitutional law of the new country was the *Aliens Act*. A proposed amendment to this Act, Article 81, stated that residents from other republics who did not apply for Slovenian citizenship would maintain legal residential status based on their permanent address or employment in Slovenia. Although approved by the Executive Council, Article 81 did not appear in the final rendition of the *Aliens Act* after it received insufficient votes from the council as a whole. Many councilors were persuaded by an argument that reciprocal agreements with other republics would take care of the legal status of non-Slovenian citizens. In the next ten years of war to the south, these agreements never materialized (Zorn 2011: 70; Šalamon 2016). The irony of the ensuing injustice to ethnic minorities is that foreigners who were not from Yugoslavia and who were living in Slovenia at the time of independence did not lose their permanent residential status. Ethnic Croats, Bosnians, Kosovans, Roma, Montenegrins, and Serbs who had lived in the country all their lives lost their legal status, while many Albanians, Turks, Italians or Romanians who were residing in Slovenia during independence maintained their legal rights. Ultimately, it is a matter of conjecture about whether this was a planned ethnic cleansing, a move to have minorities return to their republics, or a bureaucratic oversight. Come that as it may, the new Slovenian constitution did not outline provisions for persons who became aliens due to secession, and those were primarily ethnic minorities from other Yugoslav republics. According to official Slovenian records from 2009, 25,671 people lost their official legal status and subsequent social rights.

As noted in the previous chapter, Agamben's (2001: 22) state of exception in modern biopolitics is exemplified in the succession of camps from internment to concentration to extermination, which "represents a perfectly real filiation," or line of descent. Criticism is levelled at Agamben's unwillingness to distinguish between different forms of camps – or forms of exception not contained within camps – that could perhaps "dislodge the notion of a clear judicial filiation" (Owens 2009: 575). A loosening of Agamben's teleology thus enables consideration of the foundational relationship of the nation-state to other forms of exclusion. The Nazi camps were an extreme and inevitable form of a "political space in which we still live" (Agamben 2001: 36), but there is a clear and insidious line of decent to the Slovenian erasure. It is important to note that by failing to pass (by a small majority) an article that would have maintained the permanent residency status of members of other Yugoslav republics in its Alien Act of June 5, 1991, Slovenian policy-makers were presaging a particularly virulent state of exception in their nascent nation-state.

Over five and a half thousand of the erased Slovenians were under the age of 16. Historically, infants born in Slovenia of parents from other Yugoslav republics were entered into the Slovenian register of births as citizens of the father's (and sometimes the mother's) republic, but often the republics in question were not informed of the birth.[3] Moreover, to the extent that there were no legal ramifications to these designations, prior to 1991 no one was too concerned about accounting inaccuracies. When social rights and legal status in Slovenia were denied, these children were usually not claimed by other former Yugoslavian states. Most youths did not want to leave Slovenia to try to claim rights elsewhere in case they could not return home. But to stay meant the withdrawal of social rights for these children and young people including free health care, free education past elementary school, access to subsidized housing, participation state-sponsored sports, driving licenses, credit, employment (black-market employment only), and marriage certificates. In addition, as illustrated by some of the stories in this chapter, to stay in

Slovenia for many young people meant running a gamut from avoiding contact with the police and other authorities to disappearing underground. To go to a state hospital or school was to risk deportation. It is important to note, further, that the official number of erased children did not include a significant number of young people who were caught outside of Slovenia at this time and who were not allowed to return, as described with Anton's story in Chapter 5.

As the following stories indicate, lack of security and substantial privations led to tensions within families and communities. Each story of erasure is unique, often defying bureaucratic or legal rationale. What seems relatively common is that erased people were chastised and shamed, at best, and, at worst, untold numbers were deported. There was no sense of solidarity amongst erased people at the beginning of the 1990s. Many told us of ignominy and feelings of stupidity for missing the six-month window for citizenship. Others were indignant, saying that they had always enjoyed mobility and autonomy as Yugoslavs and had not intended to give that up. Still others spoke of loving their ethnic homeland and identity as Serb, Croat, Bosnian, and so forth; for them, living in Slovenia was seen as an economic opportunity and a way to better their lives and those of other family members. Amongst children and young people, a predominant feeling was one of confusion and distress at seeing their mothers and fathers under such stress. Others were confused that friends and classmates (and sometimes teachers) had turned on them, ostracizing them and calling them names. Their distress sometimes showed up as anger towards their parents.

With the stories below, I want to do two things in relation to the theoretical arguments of the previous chapters. First, I want to highlight the vagrancies and capriciousness of the erasure process in the sense that determinations were variable in space, time, and through families. I highlight examples of erased young people denied status in one administrational unit who receive it elsewhere, families who are granted legal residential status but then lose it later when they apply for citizenship, and sisters who get legal status when it is denied their brother. Second, while documenting the instability of state determinations I want to highlight how some of this plays out within family relations, and through communities. As noted above, erased children scorned by teachers or bullied by peers often turned on parents as the cause of their distress. How did this show up spatially? In what sense, were young people's relations transformed? From these empirical examples I draw parallels to the previous discussion about the good-enough parent, the good-enough state, and then push that forward to consider why acquiring sustainable ethics is preferable to advocating universal child rights.

Living erasure

> I thought about suicide, it hurt me that I was forced to buy fake, stolen documents in order to move around freely. I paid 500 German Marks for a driver's license.
>
> (Interview excerpt from Kogovšek 2010: 103)

The suicide rate in Slovenia is higher than the rest of Europe, and amongst *Izbrisani* youth it is higher still. For many *Izbrisani* young people, there was no chance of state schooling beyond the elementary/primary level; without papers there was no possibility of baptism, universal health care, playing in local and regional soccer tournaments or any state-sponsored athletics, legally driving a car or getting married. Some of these young people had grand-parents and other relatives in Bosnia, Kosovo, Serbia or Croatia with whom they had frequently visited but those visits were now difficult for fear of not being allowed to return to Slovenia. Day-to-day movement was also curtailed as *Izbrisani* youth tried to avoid police and other officials. As their struggle progressed, reporters, activists, and academics compared the plight of the Slovenian erased to the missing and disappeared in Latin America (Gregorčić 2008), and the genocide of the Holocaust (Mekina 2007, cited in Zdravković 2010). The erased youth and their families were not dead, but some argued that their situation mirrored the disappearances of Chilean youth during the Pinochet dictatorship. In many ways, others argued, the *Izbrisani* predicament was worse than death because they continued to suffer abuse and neglect.

I develop my analysis here through stories from a few young people and their families, not because their lives generalize the consequences of being erased, but because their stories are in no sense unusual. Each *Izbrisani* youth story and its relations to family, community and the state represents a unique case of denying rights, mobilities and possibilities for development through what has been politely called administrative bordering (cf. Zorn and Lipovec Čebron 2008; Jalušič and Dedič 2008; Kuhelj 2011). I begin in this chapter with the consequences for young *Izbrisani* in terms of privations and loss of human rights that suggest the inappropriateness of this term, before working through something more sustainable in the next chapter. To the degree that the loss of legal rights was a stripping to bare life, young people were parti-cularly susceptible to the machinations and whims of administrators and authority figures. At a time in life when identity formation is paramount, young *Izbrisani* were told that they were nothing.

Ana Mikšić[4] – "You are erased ... You have nothing here..."

After 1991, it was common for minority young people to lose all their docu-ments when they applied for permanent status or for Slovenian citizenship. Yugoslav passports and Slovenian identity cards were punched through with a hole-press or they were stamped as void. Astonished applicants for residential permanency watched as a local bureaucrat went through their Slovenian documents cancelling each page, and telling them to go to their country of origin to get documents. Thus began Ana's story of erasure. I met her at the space she leased for her business in a small mall on the waterfront of Koper, Slovenia's main port on the Adriatic. Ana built up her tailor business beginning with friends and acquaintances. As she was beginning this work, at 20, she and her boyfriend got pregnant. They were both Catholic and it was important that they wed, but they could not get married without documents. Remembering her 20th year, she tells me:

> I went, I had no chance for anything, getting a job or ... credit. Anything, in any direction I turned; I couldn't go to the employment bureau, nowhere. Simply, I was not there. When I went to the town hall in Koper, to ask for the birth certificate or something ... "You are erased" she said, "you have nothing here, go to Serbia to get it." I said, "How should I go to Serbia? I was born here. They don't know about me in Serbia. Just because my parents are from Serbia. I am from here."
>
> (Nov. 2013 interview)

Ana's parents are ethnically Serbian, but have lived in Slovenia since 1970 in Labor, a small village close to the Adriatic from where you can see the Croatian border. The mother worked in a nearby fish cannery and the father repaired wine barrels for a company in Koper. All the workers, he said, were Bosnian and Serbian. To this day, he has been unable to secure the pension he is due for working at the company. Ana tells me that she was 17 years old when her parents applied for citizenship and with denial the whole families' permanent resident status was revoked: "we didn't even understand what a request for citizenship would [result in]."

Ana was particularly confused, "I, for example, as a person born here, I should get it automatically, at least this is how it was said in the media." She did not understand why most of her friends got citizenship and she did not. "'Look, I got it automatically' said one, they did not even have to apply." I can still hear the resentment in Ana's voice.

Figure 4.1 Ana's tailor shop in Koper
Source: author

Classmates she had known all her life bullied Ana and her brother at school. Her father explains to me, "… the children were upset. That's it. But what could we do. We advised them that this would pass. It would be like this for some time and then it would calm down." Nevertheless, for Ana it was a traumatic time, that "got to me psychologically, in school marks, on my health … I was the only one at the school who had to stay in the corner, punished." The 'othering' is clear to her, and stated with a touch of chauvinism: "I was everything but a child – a gypsy, a Bosnian – everything except a child."

Later, Ana developed epilepsy, which she blames on her early school experiences as an erased child; "… everything together, you know. I was pretty closed and quiet … I did not want to be at school. Nobody wanted to sit next to me … there was only one schoolmate, she was also scorned just because she was fat. And we would sit together. Nobody wanted to be with us." Ana finishes with a smile but she is clearly upset. Inez says she is going to slow things down and Ana takes a deep breath.

Much of Ana's childhood trauma began in primary school, and even before the erasure she was taunted and bullied as the only Serbian at her school. Her feelings of alienation were exacerbated after 1991, but by then she was a teenager and had developed some resilience:

> In secondary school … things were happening in other ways, the hatred I felt was mostly in primary school … I thought "why do my parents have to be from [Serbia]?" Because the children usually suffer. And then I am simply thinking that we are all human beings, no matter from where.
>
> (Nov. 2013 interview)

Ana left school at 18 years of age and tried to get a job but could not because she had no documents, so she left to pursue employment opportunities in Germany. While gone, the situation with neighbors in Labor and her family became tenuous as the war to the south escalated. Ana's parents were the only Serbs in the village, and tensions increased. It came to a head when her brother got married.

Ana's brother moved to Italy when he was erased and got *permesso di soggiorno* (an Italian residency permit). He married an Italian girl on the day that the Bosnian war started, and as a consequence the Italian borders closed for some time. Ana and her parents were stuck in Italy for a while after the wedding. When there, one of their neighbors in Labor started a rumor that Ana's father was a *Četnik* and had lived in Slovenia illegally as a rebel all his life.[5] As a *Četnik* he was considered a terrorist and an enemy of the state. As soon as the border opened Ana and her parents returned home, but it was a return to a village replete with tensions and fears. Tensions are still high. When we visited Ana's mother and father a couple of weeks after I had talked with her at the Koper shop, we got lost, so we asked for directions at the local church. The gardener looked at us and for a while said nothing, then: "Oh yes, the Serbs, they live over there." I understood only a little of what came out of his mouth, but the tone and the body language registered clear disdain. Ana remembers how it felt to live here as a teenager during the early 1990s; although she did not care what people said, she had no friends and she would not walk around the village. That and all, she remembers how angry her father would get: "'Don't listen to them', I'd say to him, 'If you get angry it is for their benefit'." The tensions have lessened now, but our interaction with the church gardener and graffiti near Ana's shop suggest that they are far from over (Figure 4.2 and 4.3).

Figure 4.2 The village of Labor
Source: author

Figure 4.3 English graffiti near Ana's shop in Koper
Source: author

The arbitrariness of the erasure is clear from Ana's story. When returning from Germany, she had no trouble crossing the Slovenian border from the north and she and her parents crossed the Italian border without difficulty (while many had significant trouble crossing southern borders that joined with Balkan states; see Anton's story in Chapter 5), but trying to get documents so that she could get married was a different story. Ana was told by friends to keep trying until she found a sympathetic ear:

> Yes, I called to the ministries ... and got no answers. They were very unfriendly. So then this lady [the one from the quote above] ... they changed her with a young one ... when I was about to marry ... I was totally desperate, because I simply said: "I am pregnant, I have to marry." I went to the town hall. In the town hall they said to me, I have no papers ... They can't say anything to me, because they have to get the papers. And then I called back to Ljubljana, and I literally cried on the phone, because ... I was so desperate that I asked this girl (the young one), and she really listened to me and I asked her what can I do now. I told her that we have to marry and all, and I don't have the documents ... in December, sometime ... before New Year. And I told her that I have to marry in February and if I don't get the documents, we can't do it. I mean, because at the municipality [of Koper] they said to me that I cannot marry without the documents. And then she said: "All right." She wrote it all down and said: "Look, I will call you." And truly she called me. And said: "You can expect your..." You know, I told her the story, you know that I was born here an all ... She was asking things, I answered everything she asked and then she called me back, because she had to check some documents. She probably took the documents in her hands and checked them. She called me and said: "Look, you can expect them at the latest in January. In the first days you will get a paper document, with which you can go to the municipality to get your citizenship." And it really was like that.
>
> (Nov. 2013 interview)

Legislative inconsistencies suggest a lack of care, but Norman Cigar (1995), in an acerbic discussion of the Bosnian genocide, argues that although intransigent bureaucracies and opaque legal structures can contribute to the harassment of minorities, state leadership is the main factor contributing to ethnic cleansing, and there is always a larger constituency at work. He argues that the elite in society channel the competition or conflict among ethnic communities, and they do so at the local as well as the state level by tacitly encouraging hostility and allowing injudiciousness to go unpunished. The idea of a pure Slovenian state was the unspoken backdrop of several populist rightwing governments after 1991. In noting this, Igor Mekina (2014) suggested to me that although the erasure was originally a legitimate ploy to rid the country of Serbian military personnel it became something bigger. Tito's government required a year of military service from all Yugoslavs, and in this cross-cutting process of nation-building he required service in a republic different from that of birth. In time, many Serbian military personnel chose to live in Slovenia. It was to rid the country of this threat, argues Mekina, that the policies creating the erasure were put in place. He goes on to propose that with independence and the adoption of neo-liberal forms of governance and economy, and no backlash against the erasure in the 1990s, many Slovenian politicians saw the benefit of creating a subclass of people for whom they had no responsibility. Legislative inconsistencies and incompetency was acceptable when it came to the rights of erased people. As a consequence, the individual struggles against a state that did not seem to care helped to isolate erased people from each other, with each idiosyncratic fight against erasure seemingly unrelated to others. Nevertheless, erasure stories suggest that the heart of local processes was an intent on chasing ethnic minorities from Slovenia through intimidation, coercion, and deportation. The process pit communities against minority families, and it raised significant tensions within families.

With legal status Ana's life changed entirely, and for a while "I could scream out for joy, I mean, at that moment ... because I simply realized that everything changed for me. All life opened for me ... Literally, I am free. Wow, I am going to get married. I can get a job and all." From that time on, Ana built her tailor business beyond friends and she let go of the animosity from neighbors in the village. She does a brisk clothes repair and fabrication business, but the family's financial situation is not secure. She and her husband have two teenage children, and she worries about their future. Ana is concerned about how Slovenia is managing its economy and the growing number of unemployed youth. It seems to her that there is a precariousness to the Slovenian political and economic situation that does not bode well for her children: "Bad," she says.

> Bad here. Maybe in some other country. But here, if there won't be any changes, it is going to be difficult. Because I see, we work hard every day and all, for nothing. I mean, we barely survive and ... [The young people] have a job for a few months and then they go to the employment bureau, then they don't need you anymore. ... [My] children, and they are not small – he is going to be 18 and she 16 – are saying to me: "Mama, here it won't work ..." They can also see that there is nothing here. Because they would like to work as students, but can't get anything. There is nothing, no jobs. If you have some connections, you can get something. And it is like this everywhere in this country.

Ana is commenting on a new economic system since Slovenian independence that follows neoliberal economic principles, but is struggling with the new realities of austerity and debt recovery. Some, like Ana's parents, wax nostalgically for the days of communism and Tito when everybody had a job and a house. Nobody was rich, but nobody was poor. Ana's mother reflects on how hard it is for the aging father to be self-employed and get sufficient money for the family, "… and he is drinking more now." The parents would not survive if Ana and her husband did not contribute to their income.

Nalia Daničić – "I had a status from the beginning and [then] they took it."

Afan Daničić came to Slovenia from Bosnia in the 1980s with the guarantee of a job at the shipyards of Luka Koper. Slovenia has a 46 km coastline and at Koper there is a container port that at the time of writing rivals Amsterdam and Zeebrugge in terms of imports into continental Europe. Koper and neighboring Trieste – with its oil tanker industry – combine to provide a bustling import economy. Afan was trained as an electrician and could turn his hand to just about anything that needed to be done in the yards. He brought his wife with him to Koper and their first daughter, Nalia, was born in 1990. The family settled in Ankaran, a coastal community on a peninsula to the south of Koper, only a few miles from the Italian border. Afan applied for permanent residence status and then citizenship when Slovenia split from Yugoslavia. He received permanent resident status but his application for citizenship got lost in the bureaucratic confusion that followed the vote for independence. His legal residency status was withdrawn later at the seeming whim of a border guard, highlighting a localized capriciousness. We met with Afan and Nalia in a coffee shop in Koper.

"I came here, I was working for the common country [i.e. Slovenia] from that time till today," Afan tells me. "I applied for permanent residence and then I applied also for the citizenship. This application was there for one year

Figure 4.4 The town of Koper showing the cranes of Luka Koper and Ankaran on the peninsula just beyond
Source: author

and then they took my documents." The loss of documents occurred in 1993 when he went to Trieste in Italy on business. On returning, a border police officer confiscated his papers, and a little later, he was denied an identification card and removed from the register of permanent residents by an administrator in Koper. This had serious consequences for Afan and his family. He and his wife were registered as Bosnian and told that they had to go there to get identification cards. By this time, the Balkans war was in full force and Bosnia was a war-zone. Afan and his wife applied for and got refugee status, but when the war ended, they lost that status and, along with their children, were denied permanent resident status.

"They wanted us to go to Bosnia," Nalia explains. "They were sending us to Bosnia but we didn't have the documents. We couldn't cross the border. And the more you tried to explain [this to] them, [the] more they were claiming their position. They [just became] bigger numbskulls. They were sending us to Bosnia to go to the municipality there, because our family is from there."

Nalia and her younger sister and brother were born in Slovenia but their births were not registered in Bosnia. Sending them to Bosnia or the new Bosnian embassy in Ljubljana would have been futile because there was no record of their births in that country:

"I had a status from the beginning, right dad?" asks Nalia. "I think that I had status from the beginning and when I [was] five months old they took it."

"They demanded that she must register there and arrange all the papers but I didn't want to accept this because my children were born here," Afan notes with frustration. "We are not registered there." Nalia picks up the argument: "The more we were trying to explain, the more were they claiming their own side. Complications all the time, they always demanded such things from us."

School in Ankaran was rough, explains Nalia: "The teacher was always asking me 'Nalia, where is your health card?' ... I never wanted to speak loudly, because my schoolmates were teasing me. [One] said to me: 'oh, where do you come from!' She thought that I was born in Bosnia. She was always repeating this." The medical doctors at the clinics embarrassed her: "I mean, how to say, the doctors always humiliated us a lot. They said to me: 'Ah, again this one without the insurance' and I had to go in always as the last one. I was the fifth, in line of the alphabet, because my surname is Daničić, I am D, and they put me on Ž; they always excluded me ... they were behaving strangely towards me, although we paid everything ourselves."

Like Ana and her brother, Nalia and her brother were the only erased pupils in their school. Unlike other erased children elsewhere, they were allowed to go from primary to secondary school because Afan kept his job at Koper Luka. Most other erased people lost their jobs and had to rely on black-market employment, and their children could not go to state schools beyond primary level. For Nalia and her brother, the school teasing and beatings intensified: "They called us *čefurji* (refugee), they were calling us *čefurji* a long time." Nalia was ten at the time when this started, and she was still getting beaten up by the same girls at 23, just over a year before we met. At school, she was good at volleyball but when her team started doing really well and traveling to competitions abroad, Nalia's lack of documents and fear of deportation made it impossible for her to continue and so she stopped training. She also had to give up a graduation holiday to Greece with her classmates, but it was the teasing and beatings that hurt most.

> When I was in the secondary school I had one girl that had beaten me up a few times. She started saying this: that I am *čefur* ... that I have to go where I came from ... one time ten of them attacked me and beat me up ... Dad took the measures immediately. They were all afraid and ran away. The one that beat me, she stayed. After a few days, ten girls and four guys came in front of our house. They were knocking. I went out, I saw them and I called my dad immediately. This thing expanded and this girl stayed. I was working in [tells the name of a firm] two years ago and she was waiting for me at midnight to beat me up. Yes, two years ago.
>
> (Jan. 2014 interview)

The main perpetrator was a pupil at Nalia's school and then, later, they were both at hairdressing school and the taunting continued. Nalia stopped attending the latter school and finished her course with online exams.

Afan was able to continue working at Luka Koper, which is why Nalia and her brother could stay at school, but he was denied social housing: "I worked there for 27 years but had to now invent a variety of procedures for them to pay my salary, usually by making sure that a friend was giving out the pay packets. Sometimes I had to leave without my pay." Anger and frustration register in Afan's tone, as he points to the struggle he had against the state, which became a struggle also with his children.

> You fight for the things that then the state took it from us. I started work here years ago and I invested a lot. If I would knew that this would happen I would never ask for permanent residence. It costs me, it costs.
>
> (Jan. 2014 interview)

Nalia and Afan are on good terms now, and with EU recognition of the human rights violations incurred though erasure, the family is now united in its fight for reparations. During the early part of the 1990s, the tension for the family was palpable, and it was particularly difficult when the children were younger.

"I was five years old when my brother was born and everything started," Nalia remembers.

> I can remember that the situation in the family became, I mean ... It started. The family was full of tensions. A lot of arguing, nervousness, tensions. I couldn't explain to myself what was happening. I didn't

understand, I didn't understand why for example we have problems when we cross the border. A lot of times when we went somewhere … once, they almost kept me at the border. There was a lot of things I couldn't explain to myself, actually, I understood it only when I grew up.

(Jan. 2014 interview)

In addition, the Daničić's ties in Bosnia were completely cut. Afan, for example, was unable to leave to attend his mother's funeral for fear he may not be allowed to return. Nalia recalls:

In this way (she sighs), for example, I didn't see my relatives for my whole young life, because they are from Banja Luka. They had to come to us and when, for example, our grandfather died, I couldn't go down there. Simply, we couldn't go anywhere. Also my mother wanted to go, but they didn't let her. And it was her father. I mean, it was pretty difficult (she is at the edge of tears). And then I remember, when I was about ten, twelve years old, a lady came, later I heard she was from the social work department, and they said to my father he must arrange this, for us to get the citizenship, if not, they will … they would take the children from them (she is now crying). They would take us away (with a shaking voice).

(Jan. 2014 Ljubljana Erased Week Symposium)

It was not unusual for the authorities to put *Izbrisani* children in foster care, or to use the threat of foster care as a scare tactic to compel families to move south.

Sonja Krupić – "I hid down like a little mouse"[6]

Sonja's family comprised only her mother, who was taken from her with erasure in 1991. Her mother was working in Germany at the time and was unable to return to Slovenia. Sonja was 17 and looked after their Ljubljana apartment as best she could, working on the black-market in order to earn enough to keep paying the rent. In time she was unable to maintain the apartment and ended up on the street. Her mother stayed in Germany, where she continues to live. She and Sonja occasionally talk but they are no longer on good terms. Sonja now lives in Fužine, a poor immigrant catchment area in southeast Ljubljana, but for many years she was a homeless teenager trying to survive in the city center.

Fear without hope. Uncertainty. Hiding from police. Officials everywhere only rejected me. They had power over me. I could not cross the border. I was undocumented [for] seven-and-a-half years. I lived in fear non-stop. However, I was young. I had more courage, more nerves. I tried to be inconspicuous.

(Jan. 2014 interview)

Sonja was born in Slovenia when it was part of Yugoslavia. In 1981, following her father's death, she was looked after by her grandmother in Bosnia as her mother searched for employment. At age five, Sonja returned to stay with her mother in a subsidized apartment in Fužine. At 17, Sonja was left in charge of the apartment while her mother went to find work in Germany. In February 1992, with her mother still abroad, Sonja was identified as a citizen of a former Yugoslav republic who had not applied for Slovenian citizenship and lost her status as a permanent resident. With that, she also lost the subsidy on the apartment and decided to leave school so as to work and pay the rent. Her mother was also erased and told that she could not return from Germany.

> I never received any letter that I should arrange my citizenship. I called my mother [to see] if I have to arrange some papers. She didn't know what it was about. When I found out I was erased and they destroyed my Yugoslavian Personal ID, I was forced to go to the Bosnian embassy, but this was the last thing I wanted to do: I don't belong there, I have nothing to do down there, I grew up here, in Slovenia; here I have my friends.
>
> (Feb. 2014 interview)

As noted earlier, for *Izbrisani* the consequences of losing legal status included not just ineligibility for citizenship, but also removal from subsidized housing, and loss of legitimate employment and free health-care. Sonja tried to get help from sympathetic friends and lawyers, but she was a teenager with little experience or influence.

Figure 4.5 Fužine, a poor immigrant catchment neighborhood in Ljubljana, home to many *Izbrisani*
Source: author

I tried to arrange my status by calling to different offices, until someone started to yell at me that I should hide and be silent; if not, they are going to find me and deport me. I stopped calling and hid at home in fear that someone will call me and deport me. I hid down like a little mouse.

(Feb. 2014 interview)

For *Izbrisani* youth, the consequences of erasure showed up as strictures and rebukes, as detentions and expulsions, and as denial of access to bureaucratic processes that, as we saw earlier, seemed at the whim of low-level government officials. Sonja avoided officials or people connected to the Slovenian government, including doctors.

I tried to survive on my own. In the meantime, I was for some time without a job. They switched off my electricity. I was without money, without food. For two months, I was literally starving. I lowered myself so much that I was begging in the city center. I never felt such humiliation before! It felt so terrible that I reached this point.

(Feb. 2014 interview)

Young people are often resilient and flexible in the face of change, but only to a degree (cf. Marshall 2013). Sonja's depravation grew, and her freedom of movement was curtailed by fear of deportation to Bosnia where war with Serbia continued during the early 1990s. She got in touch with her mother in Germany to see if she could get some help: "I could not cross the border. They threatened me [with] deportation to a war zone. The horror. There I would be killed and raped. I too was raped by a friend of my mothers who was supposed to help me. [I did not report it] because I would be deported." Sonja's rape at the hands of her mother's friend created a gulf between them that has not since been bridged. To the degree that the loss of legal rights was a stripping to bare life, young *Izbrisani* were particularly susceptible to ethnic and sexual victimization of this kind.

In 2000, Sonja got temporary documentation from the Bosnian embassy in Ljubljana: "Six zeros at the end of my registration number ... Mine [used to be] 505 060 and now I have six zeros at the end. No birth-right. It was strange." The six zeros signified not only an erasure of legal rights but a denial of parents and birth, placing the young person in a position that was less-than-human, without value. If anything, perhaps the value of the *Izbrisani* youth, besides their potential as cheap labor, was that they could be blamed for the trials and tribulations through which the new country was going.

Ilmi Horváth – "I still consider myself Roma and I will stick to that until I die."[7]

Then again, as soon as I got Slovenian citizenship, things got better. I still keep pictures of places where we lived and what we had to go through and I just want to show it to people so they can see what we, as a family, had to overcome.

(Ilmi, March 2014 interview)

The Slovenian population of Roma is approximately 20,000: they mostly live in the rural areas of Prekmurje and Dolenjska, which represent respectively the most eastern and western extremes of the country. The largest urban concentration of Slovenian Roma is in Maribor, which is located in the eastern part of the country close to the Prekmurje region. In addition, there are several villages around Maribor whose inhabitants are exclusively Roma. Ilmi lives in Maribor and, perhaps because of his urban context, ideas of mobility, connection and exclusivity permeate his narrative. He remembers with nostalgia what things were like for his parents on arriving in Slovenia.

> I know this fairy tale. What happens in your life it's a fairy tale. My parents are from Kosovo. Before they came here, in the time of Yugoslavia, it wasn't so important if you were Bosnian, Slovenian or Roma, Macedonian. They accepted you. If you were working as a cleaner or miner you were a man. My parents were valued here in Maribor because they were hard working people … They went to Kosovo for holiday. They treated them as they came from a big country. Not as from Slovenia, but as they would come from Germany, Italy or America. It was like that.
>
> (Ilmi, March 2014 interview)

Like Ilmi's parents, most Roma who live in Maribor come from the former Yugoslavian town of Kosovska Mitrovica (in Kosova). Before the Balkans war more than 20,000 Roma lived in this area, but during the 1970s and 1980s they were attracted to the employment in Maribor. Maribor is the second largest city in Slovenia with nearly 100,000 inhabitants. It is the major focus of heavy industry in the country, and its location in the eastern part of the country offers important trade with Austria and Germany. In 1941 this part of Slovenia was annexed by Nazi Germany and Maribor was set up as the regional capital. Immediately after the occupation, the Germans organized mass expulsion of non-Germanic and non-Slovenian peoples to Serbia and Croatia and later there was a systematic round up of Roma who were sent to concentration and work camps in Germany.

Whereas an industrializing Maribor welcomed labor from the south, the tone changed in the 1990s, when the new neoliberal Slovenian economy declined into recession. A combination of rightwing nationalism, the European sovereign debt crisis, and a dramatic shift from communism to neoliberal state governance changed the perspective of Slovenians to local Roma: in time they got taunted as *izbrisani* (erased) and *cigan* (gypsy).

Figure 4.6 Maribor
Source: author

> Hey you, the erased! Hey you, *cigan*! Go back where you're coming from!'
> Even if we were from here ... I was two years old when we moved to Slovenia.
> All my childhood, everything I had, it's here. And they didn't recognize this.
> (Ilmi, March 2014 interview)

Ilmi describes how as a child in Maribor he felt like he was part of a normal family living in a nice apartment. What might be described as dystopia descended on Ilmi's family and other Maribor Roma when they returned to an independent Slovenia after spending the summer in Kosova. It took six months of travel through the war-torn Balkans to get back home, Ilmi was 12 years old. On returning, they found another family living in their apartment and his parents' jobs were no longer available.

> We went to the municipality and they said to us that we have no rights to get any kind of status in Maribor. From that day, were we living from day to day. We were collecting glass bottles from trashes so we could live.
> (Ilmi, March 2014 interview)

Due to the Slovenian Alien Act of 1991, which created the political process of erasure, Ilmi was barred from going to school and his father was unable to obtain health care when he contracted tuberculosis. From 13 to 23 years of age Ilmi lived in the apartments of Roma friends and acquaintances, often in deplorable conditions.

> Believe me I still see things how and where we were sleeping and living. For more than ten years... We were destroyed. You had no right, you had

troubles with every cop that stopped you, no one trusted you because you had no papers, [and] you had only a paper from municipality of Maribor where it was written that you filled in the application for citizenship .... I was walking around the market [looking for] leftovers for my brother, so we could survive. You couldn't get a job. You couldn't get support from the state. Nothing.

(Ilmi, March 2014 interview)

When Ilmi was 16, he had his first daughter, and over the next eight years he and his wife had five more children. He made a living picking out things to sell from the trash: "Glass bottles, paper, iron, everything that was useful. I was walking on the streets, collecting things. ... It was sad when I came home and my child wanted milk or food. Because you didn't have no support, no one that would help you to take care of your child. I will never forget this. It's nothing for me ... that I was erased, I can take this ... but my children were suffering too." Ilmi remembers how his hard-working father gained respect as a man and a father, while he could not even buy milk for his children.

The 1993 Slovenian Law on Local Self-government afforded Roma people specific judicial protection; in Maribor this translated to having at least one place reserved on the City Council for someone of Roma descent, but it is not entirely clear how this representation helped Roma children. In 1996, the Roma Association of Maribor was created to help deal with issues of housing and children's health and illiteracy.

"We wanted to connect with local and national institutions, to arrange our living conditions and to create a better life for today and tomorrow," Fatmir, the vice-president of the Roma Union, told us. The Roma Association of Maribor is connected with the larger Council of Roma, which serves the Republic of Slovenia (*Svet Romske Skupnosti*), which in turn is directly connected to the top of the state administration, and with the Roma Union of Slovenia (*Zveza Romov Slovenije*). Fatmir has been representing the needs of Roma for over three decades, and he is continually working on behalf of erased Roma. Illiteracy exacerbates contexts of erasure, he tells us, because many erased Roma cannot follow official documentation that offers a path to citizenship. Primary education in Slovenia is free, but there are issues such as lack of clothes and financial resources, illiteracy and weak knowledge of the Slovenian language, which prevent Roma children from attending school. That said, about 70% of Roma children in Maribor attend elementary school compared to 39% in the Prekmurje region, suggesting clear advantages for urban locations. At the local level, the Roma Union is responsible for opening access to municipal and state rights. Fatmir tells us that many Roma are able to take care of themselves today but in the 1990s and early 2000s there was a lot of discrimination. He describes how in the beginning they could not maintain the rights of children to education because of the illiteracy of Roma people.

There were problems with the people who were erased. They didn't even know that they were erased. Because of the erasure they had problems also with children. What kind of problems? If you were erased you had no right to get a status. If you hadn't got a status you couldn't sign your child to the kindergarten which is the most important thing. Even if Roma could go to primary school later on, they had big problems with language. These troubles were serious and many Roma could not finish the primary school, usually, in the third or fourth grade, they signed them to the individual school – school for disabled. But our children shouldn't go to this sort of school. They were healthy and clever. Just because of the language they had to go to this kind of school. Beside this in these kind of school children were bored, they didn't attend it regularly and parents couldn't do much about it because they didn't have status.

(Fatmir, March 2014 interview)

In 2007, Slovenia was the first European country to pass a law establishing Roma Community Councils in areas where there are a significant number of Roma people. The Roma Community Council is in charge of Roma interests, rights, cultural and international affairs but the extent of their mandate and their leverage with local authorities is not at all clear: "Even today it's not clear to me who has responsible for such acting, the state or municipality," notes Ilmi, and then adds with emphasis: "Why they were treating us like slaves? Even slaves had some rights."

Slovenian Roma are part of the current European refugee crisis to the degree that lack of education sends them abroad in search of work, but Fatmir makes the point that the erasure further exacerbated the issue of mobility.

Recently a lot of Roma is migrating outside of Slovenia. If I ask myself why: some of them lost their residential rights [with erasure] and they were evicted. For some evictions Roma are guilty because they didn't pay in time or they didn't use a chance to pay by instalments, but many of them just didn't have enough money to pay the rent. Many of them was without work and the money they received from the social service wasn't enough to pay the debts. That's why many Roma left Slovenia and in different parts of Europe they have a 100% better life …. They don't go back to the Balkan, they go to France, Belgium and German and in these countries they have better status.

(Fatmir, March 2014 interview)

Fatmir suggests that what is needed is more secure employment in Maribor, and is buoyed by a Roma themed restaurant that opened in 2014, which he says not only creates a source of employment for Roma youth but also detracts from the negative Roma imaginary.

Roma have rich culture and specific cuisine. Our food is quality because our ancestors were working on it … More important is that Roma works there and people are interested how Roma will work, what kind of food they will serve ... I introduced this point in front of the state. If there would more chances for employment [there would be] less discrimination ... Through such projects many Roma people could get a job. For example, know we open restaurant and eighteen people will be employed there, eighteen! If the state would support organizations we could evolve more projects. In this projects many Roma could be employed in civil engineering, dressmaking ... The new generations that have finished schools 1–4 and also higher degrees but they are without work and would like to work. My generation can't work anymore but this new generation likes to work.

(Fatmir, March 2014 interview)

Engin Isin (2008: 16) argues that "processes of 'globalization', 'neoliberalization' and 'post-modernization' ... produce new, if not paradoxical, subjects of law and action, new subjectivities and identities, sites of struggle and new scales of identification." This raises questions that relate to how much these new subjectivities are akin to Agamben's *homo sacer*. As noted earlier, criticism is levelled at Agamben's unwillingness to distinguish between different forms of camps – or forms of exception not contained within camps – that could perhaps dislodge the notion of a clear connection between death camps and contemporary dystopian spaces (Owens 2009: 575). Aihwa Ong (2006) argues that a loosening of Agamben's teleology enables consideration of the foundational relationship of the nation-state to other forms of exclusion. She notes that the extraordinary malleability of neoliberalism as a form of governance enables the re-engineering of political spaces and populations through exceptions beyond camps. Global market forces and the "neoliberal logic" of "emerging states," she points out, "reconfigures the territory of citizenship" through "new economic possibilities, spaces, and political constellations for governing the (national) population" (Ong 2006: 75), suggesting an important connection to Agamben and Isin's work. Ong's perspective is that neoliberal market economies help govern what Mooney (1998) refers to as disorderly people in disorderly places. Ilmi's example suggests the draconian ways that Roma youth in Maribor were disadvantaged in the 1990s and early 2000s in terms of education, health, and housing. The dystopian spaces of Maribor are part of Roma youth's circulation through low income housing and the city's wastelands. The question of new subjectivities is raised in the context of Fatmir's proclamation that a new generation of Roma are willing to work in ways that their forbearers were not. Is this a sign of hope or an example of Ong's critique of neoliberal governance?

Igor Fakaj – *"It does not matter who is really guilty, I am always blamed"*[8]

The mouthpieces of authority think that we are outcasts without value. I wish these mouthpieces could feel what we had to go through; then they would see.

(Igor, Feb. 2014 interview)

While Sonja was locked-in-place in Slovenia, others were locked-out as part of the erasure process. Fourteen-year-old Igor Fakaj was born in Germany. We interviewed Igor when he came to Ljubljana with his father and sister to seek official residency. Their case had been taken up by Amnesty International and was popularized in the left-wing press in 2014 as an ongoing abuse of rights, and in the right-wing press as another family of refugees trying to ingratiate themselves as Slovenians. 2014 marked the year when reparations for *Izbrisani* were demanded by the European Court of Human Rights in Strasbourg.

Igor's father, Amir, was deported to Albania in 1996 "simply because of my last name, I'd never been there before and have no family there" (Amir, Feb. 2014 interview). Located immediately south of the former Yugoslavia, Albania is perceived as an economically deprived, poor Balkan cousin. Amir escaped from an Albanian transitional home for foreigners and fled to Germany without documents. He was given temporary residence, and met his wife who gave birth to Igor a year later. Amir is not particularly happy with their current situation: "In Germany I live in fear. Every second the police can come and shoo me away to some municipality I have never been before."

In 2005, Amir was threatened with deportation to Kosovo ("I've never been to Kosovo") so he fled back to Slovenia with his wife and (now) four children, where he applied for asylum. Amir was granted asylum, but his children and wife were deported: "The police came and woke us up," says his 13-year-old daughter Alenka, "and they told us we had to go back to Germany. I felt very bad. I was very frightened and did not want to go back. They forced us into a car and drove us to Germany. In Germany we live in a little room where everybody sleeps. They say ugly things; kids at school say they don't wish to have foreigners and that we must leave their country." Amir's asylum case was successful and he is now a Slovenian citizen, but he is unable to return with his family. Igor sums up his feelings in a quote that pointedly gets to the heart of bare life.

> In school I am both a German and a foreigner. There are different tensions between these groups, sometimes they fight each other and somehow, it does not matter who is really guilty, I am always blamed. I feel pushed away from school, from the surroundings, from everything.
>
> (Igor, Feb. 2014 interview)

The rights of those who are nothing but human (i.e. *zoë*) are not only voided as animals, bare life also creates them as scapegoats upon which the ills of a society are heaped. Following Agamben, Rancière (2009: 132) refers to bare life as "identity fundamentalism" that takes things down to basics that can be abused, but which also sets the stage for dissent through what he calls the disruption of the sensible. Stories of that kind of disruption are part of the next chapter, but before I can get there I want to revisit the Winnicottian argument I made in the last chapter. With this revisit, I complicate Winnicott's admittedly simplistic and somewhat individualistic rendering of relations (see arguments in Chapter 3) by drawing on Rancière's notion of the sensible and its disruption. It is perhaps still a naïve and specious argument but I think is makes some sense as I move away from rights towards something that looks like sustainable ethics.

Erasure as a sustainable ethic

The stories of Ana, Nalia, Sonja, Ilmi, Alenka and Igor are of youth who lived through unprecedented trauma that resulted, in part, in a pushing back against their parents as a consequence of annihilation by the state. The stories reflect how some became estranged from their parents, at least for a while. They also reflect – and I am careful to include the parents' stories as entwined with those of the children – the ways the parents/families were estranged from the state. From a Winnicottian perspective, these children and their primary caregivers were overwhelmed through a consequential state bordering. From the perspective of Berlant's (2015) elliptical living, this was, precisely, falling apart without ceasing to exist. But for Berlant, you will remember from Chapter 2, there is something comical and whimsical about living in ellipses that is not at all evident for these *Izbrisani* youth. There is nothing whimsical about erasure. It is important to understand, then, that this form of annihilation is not the fluid and transformative process that makes up the *potentia* that Berlant, Braidotti and Winnicott preferred; at least, not yet. This and all, there is still something fanciful, impulsive, capricious and fickle about the idiosyncratic and inconsistent ways that erasure happened; the consequences then are perhaps fanciful and capricious, but for those at the receiving end they are not at all comical. Indeed they are the exact opposite: profoundly disturbing, appalling and horrific. With that said, and this is important for the arguments I make in the next chapter, although the individual contexts of erasure described in this chapter are deplorable, prior to 2002 they were precisely individuated and idiosyncratic. Each condition was isolated from others, and so it may be argued that they could not get much traction from a human rights perspective. *Izbrisani* fathers like Afan and Amir were frustrated by the vagrancies of the system but they also felt shame more than anger for what their children were going through. It was shame from what they thought was their own stupidity for not filing the appropriate papers at the appropriate places during the appropriate times. These were facts that their children reminded them of, in addition to frequent allusions to their supposed origins as enemies of the new Slovenian state.

The state legally deleted the *Izbrisani* families. From a Winnicottian perspective, they were annihilated. It is worth considering, as a provocation against the main arguments of this book, that this is where universal rights, the UNCRC, and larger pan-national legal systems come into their own. Certainly, as will become evident in the next chapter, the European Court of Human Rights in Strasbourg plays an important role as the *Izbrisani* seek collective justice. Do not universal rights proclamations hold states to account? I do not offer a defense against this incitement, but I am also concerned that at one level state diplomacy and legal processes are slippery and difficult to apprehend. Despite Mekina's (2014) legitimate concerns over a conspiracy, the official Slovenian stance to date is that the erasure was nothing more than a bureaucratic mistake. The official position is that an error was made in not including Article 81 in the *Alien Act* and that bureaucratic intolerance continued through the 1990s because of incompetencies outwith state control (but see Cigar 1995). Clearly, as will become clear in the next chapter, the Slovenian state rapidly, and particularly prior to 2004 and its petition for accession to the European Union, sought to take responsibility for the error. To date, despite a protracted process to give *Izbrisani* legal status, no government official currently in office has offered a formal apology for the erasure. What then, is the response of a good-enough state? Once annihilated, the good-enough mother elaborates a fluid border that enables a child to come back into relation differently. To the extent that the erasure process was not recognized officially by Slovenia until 2009, and not before the European Human Rights Court intervened, it may be argued that rigid bordering prevailed and continues today. Although it is entirely possible to have a valid legal system that is flexible and transformative (cf. Flax 1990, 1993; Homes et al. 2014), these practices are usually deferred in favor of something more rigid, mechanistic and hierarchical. If mothers maintain rigid borders with their children then the relationship does not reap the benefits of *potentia*. The mother fails to be good-enough, and as Winnicott (1975) documents from several of his case studies, childhood trauma, psychosis and stunted emotional development may well ensue.

For the sake of my arguments here, I want to suggest that this rigidity, this bordering, is similar to what Rancière (2009) defines as 'the sensible'. For me, bringing Rancière's work into play at this time complicates Winnicott's ideas and helps push them onto relations with the state. According to Debbie Dixon (2009: 414), Rancière's politicization of the aesthetics of the sensible is "primal" because he sees sensible space as a fundamental "*loci* for identity formation and the emergence of practices, and nodes in a regime of politics." Aesthetics, also, in this sense are primal because they come from bare life, but they also relate to Massey's (2005) ideas about countering the stasis of institutional frames when they "open space for deviations, modify the speeds, the trajectories, and the ways in which groups of people adhere to a condition" (Rancière 2009: 39). Rancière re-politicizes aesthetics away from traditional ideas of beauty and art to a consideration of the "distribution of the sensible" (Rancière 2009: 1) in terms of relations "between what people do, what they see, what they hear, and what they know" (Rancière 2010: 15–17), and where they do it. Ranciére, and Agamben before him, not only provide particular critiques of sovereign power and ways of understanding political opposition to that power, they also offer critiques of universal human rights as unenforceable and open to the variegations of state manipulation.

Pushing against the rigidity of a sensible status quo brings us to some of Braidotti's eco-philosophical aspects of the ethics of becoming, with reference to Deleuze and Guattari's project of nomadic subjectivity and sustainability, which I will say more about in later chapters. The urge that prompts my investigation at this point, then, is not only abstract but also very practical. Nomadic philosophy mobilizes affectivity and enacts the desire for in-depth transformations in the kind of subjects we become. Such in-depth changes, however, are at best demanding and at worst painful processes. These processes of change and transformation are so important and so vital and necessary, that they have to be handled with care by a good-enough state. The concept of ethical sustainability addresses these complex issues. I hope that it is now clear that we have to take the pain of erasure into account as a major incentive for, and not only an obstacle to, an ethic of change and transformation. We need also to rethink the knowing subject in terms of familial and community inter-relationality, territories, resources, locations and affects. Accordingly, nomadic ethics are not about a master theory, but rather about multiple micro-political modes of daily activism. As we shall see in the next chapter, it is essential to understand the 'active' part of activism as a push against a state that is not quite good-enough.

Chapter 2 outlined some of the contexts of universal child rights deliberation prior to WWII and as they evolved into some kinds of rigidity with the UNCRC. The unprecedented success of the UNCRC as a rights-based policy platform speaks to a number of important issues. As noted in the previous chapter, the UNCRC establishes children as subjects of rights, as having agency and as having a voice that must be listened to. This is all good for lawyers and activists, and it is good for holding the metaphorical feet of the state to the fire, but as the examples here suggest, it sometimes is of little consequence in specific locations and at particular times. The examples of erased Slovenian families point to some of the implications of public policies that pit children against parents, but also turn families against the state. For Braidotti (2013: 93–4), sustainable ethics are grounded and situated in a very specific feminist politics that I argue can be writ large in and through family relations. These relations, as I suggest in Chapter 3, elaborate a *zoë*-centered egalitarianism that allows edginess and surprise. This is not at all about human rights and freedom from tyranny, patriarchy, oppression, sexism, ageism and so forth, although these are important freedoms, but rather it is about freedom to move, to grieve, to protest, to be. Pushing this a little further, freedom for Elizabeth Grosz, is about doing; it is positive and imminent, and not contained in anything that is predictable: "it resides in the activities one undertakes that transform oneself and (a part of) the world. It is not a property or right bestowed on or removed from an individual by others but a capacity, a potentiality, to act both in accordance with one's past, as well as 'out of character', in a manner that surprises" (Grosz 2011: 72). Elshtain (1990), as noted in Chapter 3, asserts that the potential of this kind comes for children from particular and intense relations with adults. As suggested by the Slovenian stories, such relations enable children and fathers to oppose and contest societal norms as markers for what does not work in emerging neoliberal democracies. This seems to bolster my arguments from Chapter 3 that assert the importance of family relations and the authority of caregivers within a democratic, pluralistic society. The particularity of that parental authority means that it may be abused in ways more insidious than any other kind of authority, but without it an opportunity for *zoē*-centered egalitarianism would not exist. Moreover, fathering (and mothering) authority is imperative, at least for Elshtain, within an evolving democratic, pluralistic order precisely because it is not necessarily homologous with the principles of civil society. Echoing Agamben, it is pre-emptive for the organization of state power but also for the emancipation from it. Elshtain (1990) asserts that children need particular, authentic, intense and caring relations with adults. Such relations enable children and families to oppose and contest societal norms as a marker of what works in democracy. The Slovenian examples in the next chapter push these arguments a little further by highlighting some ways that young people came together in opposition to erasure policies. They feed the argument that young people are, in many ways, the most vitriolic and energetic champions of social justice, which is the centerpiece of Chapter 6.

Notes

1 I primarily use discussions with *Izbrisani* families generated between November 2013 and April 2014, although I am also indebted to interviews with journalists, filmmakers, academics, and politicians, as well as stories collected by researchers at the Ljubljana Peace Institute. Some of this work is already published in Aitken (2014a, 2016) although here I focus specifically on aspects of the stories that relate to children's rights. During the fieldwork, I was accompanied by Ines Hvala, who acted as my guide, interviewer and translator. Ines was a recent graduate from the anthropology department at the University of Ljubljana, and was well aware of the debates surrounding the erasure from her work with Professor Ursula Lipovec Čebron. In addition to translation, Ines' insight, experience and friendship were invaluable to this work. There are many other stories of the *Izbrisani* collected primarily by the Association of Erased People and Ljubljana Peace Institute to elaborate privations (cf. Zorn and Lipovec Čebron 2008; Šalamon, 2016), but our work was specifically focused on children, young people, and their relations to families and communities.

2 Igor Mekina writes for *Delo* and other Slovenian news outlets. He is the journalist who is credited with breaking the *Izbrisani* story. I interviewed him at length in March 2014, and many of the details in my account of the erasure process – including some of the conspiratorial theories – come from his insights.

3 Thirty-three countries including the USA and most South American countries define citizenship through an unrestricted *Jus Soli*, the right of soil. If you are born in the country you have citizenship rights. Another 24, including the UK, Spain, Germany and Hong Kong have restricted *Jus Soli*, which usually means that at least one parent must be a citizen. The rest, including all the former Yugoslav Republics, define citizenship through *Jus Sanguinis,* the right of blood, where one or both parents must be citizens of the state.

4 With the exception of known public figures, the names used here are pseudonyms to protect interviewees from ongoing legalities and discriminations. The fictitious surnames reflect interviewees' Serbian, Bosnian, Roma, Kosovan or Croatian origins, thus highlighting a primary gateway for discrimination and erasure.

5 The *Četnik* Detachments of the Yugoslav Army formed during WWII as colla-borators with the occupying forces, and especially with the Italians. With Serbian President Slobodan Milošević's assumption of power in 1989 various *Četnik* groups made a comeback, and Serb paramilitary groups often took on the moniker.
6 Sonja's story first appears in Aitken (2016); here and in Chapter 5 it is developed more fully and contextualized differently.
7 Ilmi's story appears in Aitken and Arpagian (2018); here it is contextualized differently.
8 Igor's story is found as part of a mobility study in Aitken (2016).

5 Youth movement rights

I don't know why in front of the zoo ... I have no idea why right there. No, it is
not that I don't know: I had this crazy idea about becoming an amoeba. And
that's how it all began.
(Aleksander Todorović, interviewed by Sara Pisotnik, July 2002,
cited in Zdravković 2010: 264)

Aleksander Todorović learned of erasure when he went to register his paternity
on his daughter's birth certificate. He submitted his identity card, which an
administrator took and summarily voided. Aleksander fought for three years
for his inscription as the father on Aleksandra's birth certificate before he
began to think about activism. His 2002 hunger strike at the Ljubljana Zoo
raised media awareness of the erased people. Prior to beginning the hunger
strike, he sent letters to local media and clergy explaining that "something
bad had happened here" (cited in Zdravković 2010: 264). At a personal level
his protest was successful. Aleksander was one of the first erased people to get
permanent resident status, and with that obtained, a re-issued birth certificate
indicated that he was Aleksandra's father. However, this was not enough; he
started the Association of the Erased Residents (DIPS, *Društva Izbrisanih
Prebivalcev Slovenije*), which became a repository for erased stories, and he
petitioned on behalf of other erased people, many of whom had heretofore not
realized that so many were in similar situations. It seems reasonable to argue,
then, that the *Izbrisani* movement began with a fathering moment, when Aleksander took issue with the absence of his name on Aleksandra's birth certificate.

In this chapter I visit more *Izbrisani* stories, but this time from the other
side of the bare-life described in Chapter 4; in this chapter I take the side that
sees *zoē* as good-enough, as more-than-human, and as something that pushes
toward radical and sustainable ethics. Empirically, I look at what happened to
erased people after 2002, when they became a movement and pushed back
against the state, forming new relations. Before I get to those stories, in what
follows I consider where these changes come from in some theoretical detail
that simultaneously embraces and thwarts us/them, citizen/erased, child/adult
borders. I begin, then, by considering the idea of authoritative bordering,
which was raised at the end of Chapter 4.

Authoritative bordering

At this point, what I hope is clear from the stories in the previous chapter that took pains to weave young people's lives in relation to caregivers, is that for young people parental relations and authority are important, and when the state tries to erase identities, familial borders are cast in high relief. I want to make clear also that I cast a wide net so as to queer caregiving/familial borders way beyond the heteronormative and as more-than-parental without dismissing what is important about the ways families are placed as an opportunity to push back against the state. I am aware that relating Winnicottian good-enough parenting to state legal and judicial processes smacks of paternalism, and I am also aware that mapping a psychological/developmental story onto state processes is a cautionary tale. Notwithstanding the resurgence of psychoanalytic understandings about how the world works (Žižek 2012; Kingsbury and Pile 2014; Proudfoot, 2015), and despite Deleuze and Guattari's (1987) concerns about pushing the structures inherent in Freud and Lacan too far, I am convinced that the externally given authority of parents/families – call it paternalism if you wish – is not necessarily a bad thing. For what I want to say here, this is not about desire as a seeming lack, and how we come into consciousness in tension with (m)others, but about the ways relations are transformative and desire is positive. The stories in the previous chapter highlight parents working through tensions with children under extreme stress, and they are also about the family as safe-haven against the excesses of the state. Still and all, it is not notions of paternalism or psychoanalysis that I want to push here, but the contexts of how relations codify, challenge and transform. In the stories that make up this chapter, positive desire and transformation are more evident. I alluded in Chapter 4 to Rancière's ideas of dissent and challenging the sensible or the status quo; perhaps, then, a push against a sensible externality is much the same as a push against the mother (or father) and, from a Winnicottian perspective, an attempted symbolic matricide (or patricide) turns relations around in ways that support children better. These changing relations, if Winnicott is to be believed, are simultaneously from the inside and outside, and they are embodied. Before I look more closely at what is possible from relational changes I want to consider the ramifications of thinking about bodies, interior and exteriors.

Boundaries and embodiments

Sonia Front and Katarzyna Nowak (2010: xiii) argue that it is through the body that we distinguish exteriors and interiors and, importantly, it is a conduit to a larger body politic. What do they mean by this? As feminist geographers noted a long time ago, the body acts as the geography closest in, providing the scope and limits of touch, smell and vision, and hence a point of reference that is a litmus test for drawing borders between the interiors and exteriors. Understood from a Winnicottian perspective, insides and outsides, subjects and objects, others and me, are complexly interwoven, and the body is the most important border to codify these connections and aspects of the self. The point that Front and Nowak (2010) make is that the embodied distinctions between what we choose to include and exclude are always political, raising lasting questions of ethical and moral value. In past work, I have considered how we distinguish between interiors and exteriors and, indeed, whether we should, but in so doing there are always political ramifications (Aitken 2014b). The contemporary feminist, post-structural, and post-humanist critique of binaries suggests that we should perhaps not make those distinctions anymore, but if we do not where does that leave us? Are we left with multiple readings and a plurality of spaces, and if so, what are the political complications of this expanding relativism? What are the political ramifications of dismissing dualistic representational categories? How do representational categories affect existential and more-than-human political crises as suggested by Todorović's comment in the epigraph. If these representations come from the inside out, whose insides are of most concern? How do these show up in the current state-pushed biopolitics?

In Aitken (2014b), I spend some time thinking about how we have theorized about places of interiority, and conclude that past psychoanalytic musings are of little worth unless they make explicit connections with relations beyond what is inside. Freud's focus on the development of self through the Oedipal-stage was challenged by Lacan's mirror-stage and a linguistic/representational turn in how we understood the development of consciousness and the evolution of unconscious drives. Feminism challenged the patriarchal basis of the Oedipal- and mirror-stages as too rigorously structured around a white, male European model of human development. As we saw in Chapter 3, Braidotti (2013) elaborates this argument with a challenge to Enlightenment thinking. By so doing, feminism obviously also challenged the patriarchal structure of society. Taking this further as part of their influential *Anti-Œdipus,* Deleuze and Guattari (1983) famously connected the focus of psychoanalysis on pathologies and developmental normalization to the apparati of capitalism and the state. By so doing, they take Freud beyond individual psychoses and mother/father relations, seeing individuals as the heterogeneous aggregate of parts of social and natural machines. Perhaps most importantly for what I want to argue here, they reposition desire as positive and productive, and the unconscious as indifferent to personal and political identity. Deleuze and Guattari (1987) empty out the Oedipus, dissipating its power into a multiplicity of 'desiring machines.' The subject eviscerates, turns inside out and reconstructs as temporary 'body without organs' along lines of desire that are positive. For example, they famously evoke the mother's breast/baby's mouth as a body without organs that comes together as a temporary assemblage, which is not part of other wholes but a working body with its own purpose. Unlike Freudian and Lacanian theory, which posit desire as search for what is lacking, for Deleuze and Guattari desire is always positive. There is no lack or unconscious desire left unfulfilled.

Deleuze and Guattari (1994) also position Marxian thought as a libidinal exercise, seeing the social and political as immediately invested with desire. The entire notion of a complete monadic 'subject' with political and personal interests of its own becomes meaningless unless understood relationally, and nomadically, in terms of its movements and fluid power. Deleuze and Guattari's analytic possibilities cannot reside in a space already occupied by objects precipitated out of the state's binary machine, like man/woman, us/them, or inside/outside (Kirby 1996: 115). The ideal state of a body without organs is relational, mobile and recombinatorial, temporary, lacking restraints, moving in excess of the boundaries and bordered authority of subjects and objects, insides and outsides.

As appealing as this is, there are some cautions to note. Kathy Kirby (1996: 117) argues that a focus on Deleuze and Guattari and other post-structural perspectives that reorder the space of the subject "can lead subjects to disrespect the bounds of others, or to cede too much of the social territory and lose the capacity to maintain their own self-interests." The danger here, she argues, is that we change the space of the subject – the mental landscape – instead of affecting the external world, institutional politics or the state. For some, the development of post-structural critiques lose the interpretative power that Freudian and Lacanian analysis pointed at the construction of the external. As a counter, Paul Ricoeur (1981) argues that psychoanalysis is a hermeneutics of suspicion that searches for deception and thereby destabilizes reliance on reason, rationality and seemingly clear meanings. Heidi Nast (2000), for example, uses a Lacanian focus on linguistic and representational power to argue that police strategies and procedures in Chicago during the mid-20th century problematically racialized landscapes around the Oedipal and the bestial. Slavoj Žižek (2010) raises the possibility of creating hope through toppling the presence of the Lacanian big Other, which constitutes a representation of hegemonic power that lies outside, external to us, and also resides within us. To do so, he calls for radical ethical acts that simultaneously change the internal and the external (us and the Big Other). For Nast, Ricoeur, and Žižek it is reasonable to laud pyschoanalysis for its interpretative qualities in that it can gets us behind the curtain to where the autocratic patriarch is revealed in all his insecurity. The epistemological question raised by Nast's work relates to how far a focus on interiorities resonates with outward affects and, if we are interested in state politics of repression, is it possible and desirable to imbricate the representational power of those relations? If we agree that within capitalist modes of production, representation simply functions as a means of coordinating flows of power that are hierarchically arranged then perhaps a new politics of creativity (e.g. Deleuze and Guattari's (1987) schizo-politics) is required. For example, Marston and her colleagues (2005) argue for the possibility of flat ontologies that eschew hierarchical power relations. Such networks of relations and power enable fluidity, and shifting into and out of authoritarian roles. For Curti and his collaborators (2011), pushing this form of relationality further, understanding differences and wholes through heterogeneous aggregates rather than hierarchies is more rewarding than positioning individuals and their distinctions within what they see as some larger normalizing or universalizing set of theories. Heterogeneous aggregates, then, comprise the muscle of relationalities, the politics of fluidity and the transformative effects of recombinatorial power.

To the extent that a combination of psychoanalytic and feminist theories of care move our thinking of places and landscapes in a direction that embraces larger understandings of affect, emotion, desire and the unconscious, it is clearly important to question the relations between interiority and exteriority. Moreover, to the extent that Deleuzian concepts of de-territorialization, re-territorialization, folds, striations and smoothings suggest a series of crowded polarities, it is still important to understand the valence of relations between what we feel inside and what we perceive as outside. In what follows I pick up from the stories of the last chapter with tales of what happened when the *Izbrisani* became a movement, literally and politically. I show how disparate and disarticulated stories of erasure come together as a representational and relational embodiment, to affect change. The change simultaneously destabilizes the state and foments change for the *Izbrisani* by elaborating a community of care through affective citizenry (cf. Askins 2016; Arpagian and Aitken 2018). The argument I want to get to by the end of this chapter is that social justice is often best constituted by processes of emotional citizenry, which are embodied and "embedded in the complexities of places, loves and feelings," and which are "beyond claims to and exclusions from nation-statehood" (Askins 2016: 515). For *Izbrisani* young people, this reformulation and repositioning enables a political force that was good-enough to push back against state strictures.

The *Izbrisani* movement

Aleksander Todorović is now widely recognized as the initiator and most important beginning activist of, and contributor to, the *Izbrisani* movement. In February 2002, he began a hunger strike at the Ljubljana Zoo, where in "becoming an amoeba" (see epigraph) he chained himself to a tree. At his wits end after ten years of erasure, everything came to a head when his paternity did not show up on his daughter Aleksandra's birth certificate. His allusion to becoming protozoa suggests a powerful political force that is encapsulated well by Agamben's (1995) embodied notion of bare life and it also resonates with Deleuze and Guattari's (1987) body without organs. It is from this latter context – Todorović's becoming amoeba – that a political push against exterior political forces is enabled. By so doing, whether consciously or not, he was pushing against state biopolitics. Agamben's work dovetails nicely with Deleuze and Guattari because it latterly offers the possibility of radical change. Simultaneously "excluding bare life from and capturing it within the political order," according to Agamben (1995: 9), makes state political processes vulnerable because "the state of exception ... in its very separateness [becomes] the hidden foundation on which the entire political system" rests. This is important for when sovereignty loses its precision and authoritative borders blurs: "the bare life that dwelt there frees itself ... and becomes both subject and object of the conflicts of the political order, the one place for both the organization of State power and emancipation from it." In the stories of *Izbrisani* bare life that follow in this chapter, there are movements away from the privation, misery and tension witnessed in the previous chapter. There is the beginning of hope, mobility, and the fomenting of a collective biopolitical remonstration leading to the development of communities of care and emotional citizenry (Askins 2016).

Todorović's hunger strike in 2002 opened space for deviations away from neoliberalism as exception (Ong 2006, 3), towards a flourishing of *zoē* as good-enough (Braidotti 2013). His formation of the DIPS not only provided a political focus for the group, it also created a forum for the collection of erasure stories (cf. Zorn and Lipovec Čebron 2008). Ismeta was 20 when she heard about other erased people.

> One day on TV I heard about the Association of the Erased Residents (DIPS). I had no idea that there were so many erased people. Oh dear, when we started to talk ... I felt like I got wings. As if a stone fell from my heart. Pains literally began to peel off my body. I could feel life, health, the future. I could see the light at the end of the tunnel.
>
> (excerpted from Zorn 2010: 23)

The *Izbrisani* movement was about coming out of obscurity and into the light. In February 2003, the first 'Erased Week' was organized to bring public awareness to the plight of the *Izbrisani*. Demonstrations and solidarity showed up on the streets of Ljubljana. The *Izbrisani* movement was not only about the privations endured by children and young people, it was about mobility and movement for a people who had stayed put, not wishing to be seen moving around cities or the country.

A series of hunger strikes in February 2005 began at the Croatian border and then moved to the UNICEF headquarters in Ljubljana. The joint themes of movement, mobility and presence motivated a large part of the agitation. Mobility within Slovenia's borders focused attention on the implicit and explicit immobilities of erased people, and especially the women and children. Movement in this instance is about the erasure of interior/exterior borders and forced sedentarism. "I've got some of my self-confidence back," Marjana, a friend of the Todorović family, tells us, "I don't wait around stuck in a corner any more. These travels around the world as an activist made me pick myself up. I say to myself: I'm not crap." In 2006, a group of *Izbrisani* activists (including Marjana, Aleksander's daughter Aleksandra and several other young people) marched from Piran, on the coast of the Adriatic through Postojna and on to Ljubljana. "Actually we did not march between towns, we drove" says Andro Duvnjak (we'll meet his son in a moment). "We went like

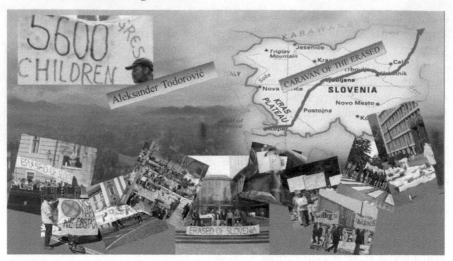

Figure 5.1 The Erased Movement, 2002–2012
Source: created and compiled by author

this, … for some time walking for some time driving. Where the roads were without settlements and towns, we were driving. And then we went through towns like Postojna and Logatec. We went by foot and came walking to Ljubljana. Everything was photographed. We were in every town. It was on TV." The march was dubbed 'The Caravan of the Erased' (see Figure 5.1). A little later, and with several lawyers in tow to aid the case, the caravan drove to the European Court of Human Rights in Strasbourg. The petition to the European Court was successful as Slovenia had just acceded to the European Union, and there was now push back. Reflecting on the case, Boštjan Zupančić (2016: 14), Professor of Law and Judge at the European Court of Human Rights in Strasbourg, famously referred to the Slovenian government's insistence that the erasure was nothing more than a bureaucratic error as "a clerical attempt at 'ethnic cleansing'." With this validation, the *Izbrisani* gained representational power and motivation to further provoke Slovenian responsiveness.

Young *Izbrisani* were joining each other and rallying to disturb what was the status quo, the sensible. A sustainable ethic is achievable through joining (cf. Askins 2016), but for rights to be attained, disruption is required. Rancière (2004: 302) famously points out that "the Rights of Man are the rights of those who have not the rights that they have and have the rights that they have not," implying that the rights of those who are nothing but human (i.e. *zoē*) are simply void without dissent. As noted at the end of Chapter 4, Rancière (2009: 132) refers to bare life as "identity fundamentalism" and argues that what is needed is the staging of dissent, a disturbing of the sensible, a dispute over what is given. Following Agamben, Rancière (2004) notes that stateless people are able to stage dissent and demand their rights and challenge the status quo in ways that are impossible for those who are part of *bios* (cf. Parekj 2014). Similarly to Laclau's (1990) notion of dislocation and de Certeau's (1984) notion of surprise, Rancière argues that real and appropriate politics occurs when there is a disruption of a hegemonic or dominant mapping of the sensible. For the *Izbrisani* there was a radical disruption of identity as a consequence of bare life that began in 1991 and led to a state of exception through which they were erased. The bare life of the *Izbrisani* found political form and witnesses with Todorović's initial hunger strike, which Žižek (2014) would describe as an event in transit, speaking truth to power, to the degree that the *Izbrisani* case became a movement. The power of movement and dissensus, of speaking truth to power, comes across most poignantly in *Izbrisani* youth stories below, the first two of which are continuations from those begun in Chapter 4.

Nalia Daničić (redux) – "I understood it only when I grew up."

Nalia Daničić's story is replete with family tensions and misunderstandings from her youth in Ankoran. As we saw in the last chapter, she had trouble at school, and her father intervened on more than one occasion when bullies hauled up at her doorstep. In 2014, Nalia participated in a symposium that we organized at the University of Ljubljana as part of the 11th erased week. We purposefully focused on the children of the erased, and several reporters showed up as part of the audience. Later, the stories of the young people who were part of the symposium were published in Slovenia's most popular newspaper, *Delo*. Irena Pan for POP TV (a Slovenian television show similar to the US's 60 Minutes) also interviewed some of the young people. At the symposium Nalia spoke of the privations and tensions I raised with her story in the last chapter. Her father, Afan, whom we also met in Chapter 4, was in the audience and when Nalia got emotional at the part of her story where she was almost taken into foster-care, he is clearly distressed:

> … and then I remember, when I was about ten or twelve years old, a lady came – later I heard she was from the social work – said to my father he must arrange this, for us to get the citizenship, if not, they will … they would take the children from them (now she is crying). They would take us away (with a shaking voice).
>
> (Excerpt from Nalia's presentation at the Child *Izbrisani* symposium Feb. 2014)

It took five more years for Nalia to get her permanent residence arrangements status, and she still does not have citizenship. This notwithstanding, she cannot be deported because Afan's citizenship gives the family legitimacy. He and Nalia are now actively working with DIPs to push the Slovenian government on reparations, as mandated by the European Court of Human Rights in Strasbourg in 2012. Part of the lawsuit on their behalf argues that a large percentage of people like Afan who applied for citizenship during the six-month window were denied solely on the whim and capriciousness of local administrators and there was no systematic attempt to stop this injustice.

Sonja Krupić (redux) – "I won and now I try to live on."

Sonja Krupić's continues to live in Fužine, the neighborhood where she was abandoned by her mother during the erasure. Recovery from the experience of erasure was difficult for Sonja.

> It was ... difficult. Until this I only try to survive. When I got citizenship then I got ill, ill. Not mentally ill, but it was difficult. When I got citizenship then I got ill, ill. Post-trauma, yeah. Like military, yes posttraumatic stress syndrome. It all [had to] go out, you know?
>
> (Feb. 2014 interview)

Sonja was involved with DIPS for a while, and contributed her story to their push for legitimacy. She is nonetheless very critical of politicians and activists alike: "I'm not a member of any party and never want to be: for me, all forms are small-minded customers." For the most part, she prefers to work on what she considered as her internal issues through reading and art: "I read books, you know. I don't know. What kind? Louise Hay *The Life is Yours*, [1] if you know it. Books like that. You know? Some are spiritual ... I try to help myself, you know. I have really great friends who never judged me ... It's not a problem. But at the end you are alone. You have to deal with yourself."

Today Sonja works as a caregiver at a local home for elderly people. She sustains her healing from the trauma of the erasure – her radical ethical act is through painting, which uses art as therapy for the trauma she went through. She shares her art with the elder community she works with, sometimes hanging her work on the walls. Sonja's paintings are stark, even brutal, portrayals of placid mountain and rural scenes.

One of her more telling pieces is of Slovenia's iconic Triglav Mountain, meaning simply three peaks. It is an ironic painting in its baldness because Triglav was used by the *Izbrisani* movement to appropriate the 'I FEEL S(LOV) ENIA' tourism campaign. The usurping was about changing the three peaks of Triglav into three hooded figures resembling the Klu Klux Klan (see Figure 5.2).

"I only try to survive," Sonja tells us. "I am sometimes angry; lies on media and some people believe them. I don't need that the government to say 'sorry', but don't tell lies ... Now I wish only that they live me in peace. I know that this was a period of my life and I can't change anything in the past. It was hard and I won't forget it. I won and now I try to live on."

Franjo Duvnjak – "I wanted to see how France looks ... and I decided to go."

Franjo Duvnjak (Andro's son) and Aleksandra Todorović were two of the young *Izbrisani* who went to the European Court of Human Rights in

Figure 5.2 Sonja and art therapy
Source: created and compiled by author

Strasbourg with the Caravan of the Erased. Like Aleksandra, it was Franjo's father who got him involved in the mobility and activism, with the first Caravan of the Erased (see Figure 5.1). Franjo was 14 when he joined the second Caravan of the Erased, and the trip to Strasbourg was his first time out of the country. Franjo's mother, Mirna, was also erased and became an activist with DIPs in the early 2000s when he was still in elementary school. Mirna was born in Čakovec, a border town in northern Croatia that abuts the Slovenian border. Franjo is candid that his trip to Strasbourg was mostly holiday: "I wanted to see how France looks like. I have never been so far, just in Slovenia. And I decided to go."

Franjo focuses on the fun aspects of his trip to Strasbourg but, importantly, later in the interview his mother makes a pithy comment on some poignant and more subtle affects: "Like you have something and that something gives you energy. And this persistence I have inside, I move on that" (Mirna Duvnjak, Feb. 10,2014). The question of significance here is the degree to which internal energy of this kind can disrupt the status quo and change abusive regimes of power. Perhaps I am making too much of this, but it seems to me that regimes of power and oppression can be changed through smaller, playful disruptions of this kind. Moreover, as Henricks (2015) points out, and as suggested in my discussion of play theory in Chapter 3, the kind of playful energy that Mirna and Franjo allude to is also the kind of energy that creates community. Franjo is candid with us that the trip to Strasbourg was as much a holiday for him and the visit to the European Court was a lark: "I was sitting on the bench at the back. I wasn't even listening. I was there … just like this (gestures boredom)." We laugh at his gesture but it is nonetheless suggestive that Franjo's lark and Mirna's activism join together in a community of care on an emotionally charged journey that played out for mother and son in different ways

Sandy Marshall (2013) argues that Rancière's relational and radical political aesthetics reside most potently amongst young people, and that protest could arise from less structured and more relational, serendipitous, playful and mobile processes. Play clearly creates community (Henricks 2015), but it is also an important part of protest as suggested by my discussion in Chapter 3 (see also Woodyer 2012). In the geographic literature, young people's spontaneous disruption of the sensible can be traced in part to Katz's (2004, 2011) work on radical play and the way children use things dispensed with by adults (cracked bricks, broken blocks and other bric-a-brac) creatively and spontaneously. Her focus comes from Walter Benjamin's (1978) idea that children's play is mimetic not just in the sense of copying something but also as a radical flash of inspiration and creativity when something is performed or used differently (see also Aitken 2001). For Katz, play is identity making and can also be revolutionary and world-making (cf. Woodyer 2012). In play, children learn and toy with the meanings and practices of their social worlds, but it is not just about social reproduction and community building; as Benjamin reminds, it is also where received meanings and relations are refused or reworked (Katz 2011: 56). The aesthetic created by these spaces and the practices of children and young people that pass through them suggests not only dislocation and surprise, but suspension of the sensible. In this sense, Marshall (2013: 54) uses Rancière to highlight "how we might understand the present political moment through the lives of children and how children play a role in building alternative futures ... how children both perform and transform the aesthetics of suffering." For Marshall, young people's disruptions of the sensible show up in their art and play, but I want to argue through Kallio and Häkli's (2011, 2013) ideas of children's politics in the day-to-day to Staeheli and her colleagues' (2013, 2016) forceful declaration that there is not only radical aesthetics in play but also an activism that can change the world.

Lynn Staeheli's project is to engage with young people and their support networks in areas of conflict (Staeheli et al. 2013, 2016). Focusing empirically on youth work in Lebanon, Croatia and South Africa, she and her research team engage the radical relationalities through which young people push back against the state through play, community, memory, and technology. Part of their project is to work through the day-to-day routines of young people and youth-based organizations to get at something larger. One paper, for example, elaborates how young people meld with urban spaces in Beirut and how those spaces in turn are changed by their presence and memorialization (Staeheli and Nagel 2018). In other studies, she uses youth activities and youth communities/organizations to focus on civic participation, activism, and learning citizenship (Staeheli and Jeffrey 2017), or how ideas of nationhood are learned in divided societies through institutional structures (Staeheli and Hammett 2013). The important point of Staeheli's larger project for what I want to say here, and I will return to her work in the next chapter in a slightly different context, is that change foments from radical relationalities seen in young people's day-to-day politics and technologies, communities and memories. This is perhaps best exemplified in the continuing day-to-day struggle and work of Aleksander Todorović's daughter, Aleksandra.

Aleksandra Todorović – "I cannot replace him. But I will work on this"

Aleksandra remembers the day when her father returned by train from Ljubljana to Ptuj after the first hunger strike.

> Mom and I walked towards him; I ran to him and hugged him. For a few days, I wouldn't let go of him. At that time, he told me a lot, but I don't remember anything. Mom told me that dad is fighting for human rights and that he is doing good. I was proud of him.
>
> (excerpted from MM RTV Slovenia, Feb. 2014)

She recalls this as a time of tension in her family because her father was dubbed an 'enemy-of-the-people' by the rightwing press: "[Dad] never wanted to confess how much it hurt him, how much some things got to him. It is not a secret that he was a very depressed person. He had dark periods when he closed himself in the apartment and wouldn't go anywhere."

In what follows, I attempt to say something about the tensions, contradictions, and dilemmas suggested by Aleksandra's memory of her father in the above interview excerpts. Parents who became active with DIPS often were pushed into the media limelight, especially if they had remarkable stories of erasure. At times, and for some children, parents were heroes; at other times, or for other children, they were enemies of the state. Feelings of despair and shame, of depression and denial, for some, gave way to reconciliation, hope and confidence as the *Izbrisani* movement progressed. In time, with activism for some and insurgent, emotional citizenship for others, there came to fruition what I am calling communities of care. Although the stories are not always encouraging, I argue that positive outcomes are possible from even the most tragic events.

Aleksander Todorović died the week before I was going to interview him. He killed himself in a motel just outside his hometown, Ptuj. He had been on suicide watch in a hospital several years previously, and so his disappearance a few weeks prior to the suicide was not a surprise to those who knew him. Aleksander had been recently beaten up by rightwing nationalists, and some wondered if the disappearance was an abduction. When his body was found, most equated the suicide to his depression. Aleksandra took part in our Child *Izbrisani* symposium at Ljubljana University in February 2014, only two weeks after her father's death. A week later I met with her and a family friend, Marjana, in Ptuj, to talk about the *Izbrisani* movement.

Marjana remembers the toll of erasure disrupting normal day-to-day life: "Yea, life was not normal, this is… this is a fact. … The family – it took its toll. If your mind does not go crazy, it all creeps up on your other organs … all that feeling of not belonging. Anywhere. It all leaves consequences." For Aleksandra, life evolved into a weird concoction of normality, disruption, external interests and internal tensions. She tell us that life seemed pretty normal at the beginning of the *Izbrisani* movement: "Mom was at work, father was taking care of me. He would pick me up from kindergarten and he would take me skiing, to spa. He taught me how to ride bicycle … He was my Daddy. If he was at home, everything was alright" (excerpted from MM RTV Slovenia, 2014). As the DIPS movement gained credibility, the Todorović family began to implode: "When I started to understand things and my classmates started seeing [dad] on TV, people turned away from him, and I wasn't so happy any more. More than once I asked him to stop" (excerpted from MM RTV Slovenia, 2014).

Aleksandra was six years old in February 2003, when the first 'Erased Week' was organized to bring public awareness to the plight of the *Izbrisani*. Three years later, in February 2006, she was with her father as part of the thwarted attempt to occupy the National Assembly, which turned into a street protest that blocked major roads in Ljubljana (see Figure 5.1): "Mmm yes, I went with him on [those] demonstrations," Aleksandra recalls. "I don't know much from the beginning when I was still little, but at about nine years old [I remember] bodies lying in the street in white jump suits spelling out the name *Izbrisani*" (February interview, 2014).

The next several years were hard on the family. Aleksandra remembers that she and her father argued most during her teenage years. She recalls specific times when she was embarrassed about his appearance on television. She wanted to tidy him up: "I used to buy him sweaters, but he couldn't be told anything … Nothing else mattered to him, just this, he was seeing just this, this was like his mission" (excerpted from MM RTV Slovenia, 2014). Tension at home arose, Aleksandra remembers, when her father started focusing exclusively on the plight of the erased: "In some ways he neglected us, so mom didn't have such a strong wish to participate any more. And she didn't want to think about it anymore. She wanted peace. But she never stopped supporting him" (Feb. 2014 interview). Aleksandra recollects arguments: "[Dad] was walking behind me in the apartment and locked the doors so that I couldn't go out and – I don't know, he was really terrible sometimes." Her parents argued a lot also, she tells us, even though her Mom always supported her husband: "He didn't even know how to use the keyboard of the computer, so he wrote on a piece of paper … or something, and then she transcribed it on the computer, taught him. Also when he needed money, she gave it to him. I don't know, she was always standing by his side, but then she had enough and he had to continue on his own. But she never stopped helping him" (Feb. 2014 interview).

In her late teens and early twenties, Aleksandra started to understand the erasure more fully; she recalls that she and her father started getting close again: "He would inform me about everything. He was the most proud when the verdict of the European Court of Human Rights arrived. He was jumping around the apartment shouting: 'We won!' I too was happy; I didn't feel ashamed any more. I became proud of him."

Aleksandra is sad and angry about the lack of any formal resolution to the *Izbrisani* issue, and she is determined to continue her father's work. In addition to fighting for reparations, she wants to see the history of the erasure in school textbooks. Adding to this sentiment, Marjana says that she "hopes that we raise our children in the way that they know what is right and what is not right. That they know what the human rights are, and they won't ever again allow such violations to happen to them. That they will be able to stand strong for their rights and that they are going to be self-confident in their struggle. That they are going to know that politics shouldn't influence their lives by violating their basic human rights. This is important." Aleksandra thinks about this and then adds something that suggests a shrewd understanding of young people in Slovenia and, as I will argue in a moment, a move towards something that closely resembles sustainable ethics.

> ... I can see in my [peers] that they are against the violation of human rights, but they would never fight against it. They would never expose themselves, because they know that this can lead to bad things. For example, especially the children of the erased know what this all means, but the society is not ready for this yet. It blindly follows some instructions and lies from politics. It is easy to support something quietly, but when it comes to doing something, it is a completely different story. Here we should achieve more. We lack open people, we really really lack this, so I don't know, it is about your attitude and if you are doing something good, this can't be something bad. If you are a public person or you are not a public person, so I don't know, here they perceive this, when you are doing something against politics, as something bad. This is something that needs change.
>
> (Feb. 2014 interview)

Aleksandra also understands that her father's fight for the *Izbrisani* and his activism not only made him unemployable but, ultimately, took everything that he had. She does not intend to go down that path.

> Yes, yes, in my opinion it is right, although I could never replace him, he was special, he didn't care at all, at the end, what somebody is thinking. Just that he values his rights and that he does what he wants to do. I cannot be like this, there is an entire life in front of me. I will do what I can in order that something is done on this case and that people at least talk about it a little, I will struggle also for the textbooks. I cannot replace him. But I will work on this.
>
> (Feb. 2014 interview)

The last time I was in touch with Aleksandra she was planning to enroll at University of Ljubljana to study, fittingly, social work, biology, and political science. Without knowing details of the University's curriculum but knowing the passions of several faculty members in these departments, it seems to me that these three subjects are the perfect conduit to help Aleksandra elaborate the relations communities of care and affective citizenship as a push against the excesses of bio-politics.

Bio-politics, for how I want to use these stories, are about the molecular and how it envelopes state and family relations. The toll on parents is palpable, and as suggested by a juxtaposition of Marjana's and Mirna's reflections with what is said by Nalia, Franjo and Aleksandra, there is often wisdom gained in the generation that follows. Even though the rights of those who are nothing but human (i.e. *zoē*) are voided, and bare-life also creates them as scapegoats upon which the ills of a society are heaped, there is nonetheless hope that derives when people get together to form spaces for insurgent citizenship, which is precisely what citizenship should do, i.e. its affects. That this arises from bio-politics and familial emotions is clear from the Todorović family's story. This is the epitome of the sustainable ethics as I describe them in Chapter 1, which is in the first instance about opening the possibility to be 'in' something, and second, it is about fulfilling life to the fullest now, rather than preserving the status quo into the future. Sustainable ethics are about change and becoming. I think that it makes some sense from what I suggest in Chapters 2 and 3 to be suspicious of universal human rights, which may be important hooks for lawyers to hang arguments upon, but are nonetheless questionable when it comes to everyday living. Instead I guardedly move towards something that looks like sustainable ethics that are locatable in and through families and communities of care, but there is nonetheless more than one cautionary tale from the *Izbrisani* youth stories that gives me pause, with an important reconsideration of the effects of movements, borders and closures. Anton's story provides one of those cautions; his is a move away from families and care to mobilities and insurgency.

Anton Horvat– *"Who are you to decide whether I come home or not?"*

Anton's story is different from those described above. It begins with mobility and suffering, but takes place outside of Slovenia's borders. Anton was 18 years of age when Slovenia gained its independence. He was born in Germany and spent the bulk of his childhood in Skofja Loka, a small town to the northwest of Ljubljana.

Anton remembers "a fantastic childhood ... in woodland all the time. I was born in Germany [but] the only place I lived was Slovenia, and I always thought I was Slovenian. I went to school there, I was a school chess champion, and third in the country." Born to a Muslim father and a Christian mother who are both Croatian, Anton lived through boyhood and his early teens in Slovenia. As a young boy Anton was patriotic: he represented Yugoslavia at the youth world chess championship, and joined the country's youth brigade, where amongst other things he learnt how to live off the land (a skill he would later draw on to survive during the wars in Croatia and Bosnia). At 16 years of age, he relished the required four month's basic military training with the Yugoslav National Army.

In 1991, when war broke out, Anton was on vacation at his grandparents' home in Croatia. His grandparents are Catholic Croatians, but their village is mostly populated by people who are ethnically Serbian. With the enactment of the *Alien Act* and without an application for citizenship, Anton was unable to cross the Slovenian border. Anton was locked out of what he considered his home country and on returning to Croatia he was unable to reach his grandparents because of the Serbian presence. This was June 1991 and the YDA (Yugoslav Army) had just entered Croatia. Unable to get to his grandparents, whose village was surrounded by the YDA, Anton joined the ZMG (the Croatian Territorial Army) and started fighting. Two years later he switched to the Bosnian army and fought against some of the worst that Milošević's troops could offer; within two years Anton was living off the land and although he says he became skilled at survival, he says he was "hungry and living like an animal."

Anton got identity documents from Bosnia, which designated his birth country and parents with six zeros (000000) in place of names because, like Nalia, he was not registered as a Bosnian citizen or in Ljubljana (see Figure 5.3). In 1995, Anton got a student visa for New Zealand, where he studied for a few years. He later got refugee status and asylum in the UK and moved there, where he met his wife and had a son. Currently, Anton has a well-respected job as a chef

Figure 5.3 Anton Horvat's erasure
Source: created and compiled by author

in London, but his refugee status is now revoked and he is on a temporary work visa. He returned to Slovenia to visit his hometown for the first time in 2000, and encountered some considerable difficulty crossing the border.

> In 2000 [my lawyer brother] was with me and so we made a complaint when we were turned away. We left, and it wasn't easy. We went to [a] different border. We turned around and then we drove to Metlika. And uh, we slept over there, in Rogatec, and then next morning we made, uh, a move to another border. Uh, close to Brežice.
>
> (March 2014 interview)

At Brežice they were successfully processed because a border guard was willing to recognize the legitimacy of his Bosnian papers. With that day began Anton's struggle to secure legal residency status in Slovenia, a struggle that was ongoing when I met him in Ljubljana in 2014.

Anton is resourceful; his military training in Croatia and Bosnia enables him to move around undetected: "I was trained as a dog-of-war," he tells me, "and I know how to live off the land and cross borders undetected." More than anything, he wants to return to the land of his youth. Anton famously and publically stated to a Slovenian politician that "Neither you nor your government will stop me coming home as far as I'm concerned. I'll come here as I please and who are you to decide whether I come home or not?" (Related to me in interview March 17, 2014 and in a Pop TV (Slovenia) documentary a week later.)

Anton's story is about an individual struggle that is not necessarily connected to family, community or official activist movements in Slovenia. He is willing to have his story parleyed as part of DIPS, but like Sonja he is not too concerned about joining the organization. Anton's story does not mesh well with the ideas of emotional citizenry and communities of care that seem to contextualize the other *Izbrisani* stories in this chapter. Rather, and more so than the other stories articulated here, Anton's is about mobility, freedom of association, and unfettered connection to the land of his youth. His story resonates in complicated ways with the example of Faroe Island culture with which I began the book. It contrasts with the romantic, bucolic, peaceful and serene idea of the moving-child-in-nature towards a more radical (and disturbing) connectivity. In arguing for understanding alternative ways that children come to know the world, Peter Kraftl (2013b: 3–4) deftly points out that children grow through and with "multiple spatialities of alternative education" and contexts for growth that include "a range of political and philosophical convictions ... that have gained increased currency, especially within anti-capitalist movements, intentional communities, squatter camps and local community economies." What Anton's story does for me, while not necessarily connecting to the communities of care that are engaged in other *Izbrisani* stories, nonetheless connects to an insurgent citizenship and a self-reliant mobility that flies in the face of authoritative borders and hierarchies of power.

Mobility, activism and locatable feminist ethics

This chapter elaborates young people's activism and mobilities, but it is also a continuation of the previous chapter's arguments for reconsidering young people not just as objects and subjects of rights, but as insurgent and affective citizens connected in some cases to families and communities of care and, in other cases, simply to insurgency. Theoretically, the chapter draws on Bergson's (1903) work on movement and mobility that was discussed in Chapter 3, which it elaborates through Braidotti's (2013) idea of being fully 'in' something, and about fulfilling life to the fullest life as it moves forward. Also from Chapter 3, I bring Rancière's (2009, 2015) ideas about pushing against the sensible, the status quo, and the framings of authoritative bordering, as part of what might be thought of as the revolutionary aspects of play and creating communities of care. Along similar lines, as noted in the previous chapter, Grosz (2011) suggests a Braidottian new feminist locatable politics, and the importance of a Bergsonian/Deleuzian "freedom to" (move, work, participate, demonstrate) in addition to the older feminist "freedom from" (abuse, patriarchy, imperialism, the excesses of capitalism). For Rancière, Grosz and Braidotti, this form of freedom is an action, a doing, and a new way to practice ethics of care that are locatable, but not necessarily in the sense that they happen some place. The context of locatable feminist ethics, rather, is the freedom to be fully in something differently and thereby, perhaps, to start fulfilling life to the fullest. When this happens in the perpetually unfolding moment of the present then it suggests as much as can be said about sustainability and potential.

Stories from the Slovenian erasure in this and the previous chapter provide examples of the loss of legal status on young people's mobilities and identity, and its effects on education, health, and familial and community cohesion. This chapter also elaborates some of the ways young Slovenians who are/were locked-in-place began to push against these strictures through reclamations of mobilities and a re-territorialization of the term 'erasure' through the politicization of the term *Izbrisani* and the creation of a youth social movement. These two empirical chapters, then, begin with ideas of bare-life and *zoē* politics and elaborate relations to youth movement and mobility to end with youth activism, social movements, and the possibility of sustainable ethics. This sets the stage for the next chapter, which uses the Slovenian erasure as a springboard for discussion of youth movements elsewhere.

Note

1 Louise L. Hay: *Heal your Body A-Z: The Mental Causes for Physical Illness and the Way to Overcome Them.* The book is translated in Slovene as *Life is Yours.*

6 Re-thinking the presence and protests of young people

> Another word for Spinoza's conatus is self-preservation, not in the liberal indi-
> vidualistic sense of the term, but rather as the actualization of one's essence,
> that is to say of one's ontological drive to become. This is not an automatic, nor
> an intrinsically harmonious process, in so far as it involves inter-connection with
> other forces and consequently also conflicts and clashes. Violence, pain and a
> touch of cruelty are part of this process. Negotiations have to occur as stepping
> stones to sustainable flows of becoming. The bodily self's interaction with his/her
> environment can either increase or decrease that body's conatus or potentia.
>
> (Braidotti 2006: 139)

At first blush, and as discussed in the run up to this chapter, Braidotti's use of
Spinoza's ethics, and particular her evocation of *conatus* and *potentia*, as
being 'in' something while simultaneously fulfilling life to its full respectively,
suggests a positive and beneficial process. And to the degree that it is possible
to become something other, something different and affirming, then yes, it is
beneficial. Pairing Braidotti's use of Spinozan ethics with Winnicott's idea of
potential spaces as I did in Chapter 3 also suggests the idea of a playful, safe
space of becoming in relations with a good-enough caretaker (whether that is
mother, father, community, or the state). This smacks of utopian thinking,
which is not at all bad to the degree that it generates hope (Harvey 2000).
Nevertheless, life is neither just nor fair, and it is never utopian. The process
of becoming is not always, nor should it be, necessarily smooth or without
tension. The previous two chapters describe how this process foments in
gruelling and deprecating ways for young people in relation to a diminution
of rights. *Conatus,* then, is also about striving and self-preservation, which is
brought to highest relief when relations with the so-called caretaker fail to be
good-enough. Winnicottian annihilation provokes new relations, and is painted in positive terms if the object of annihilation is good-enough. Winnicott's
ideas, then, are about desire for new relations and I want to emphasize that
this kind of desire, from a Deleuzian reading of Spinoza as noted in Chapter 5,
is always positive.

The last couple of chapters articulate very poor relations between the Slovenian state and erased people, but Alexander Todorivić's moment at Ljubljana

Zoo propelled change, differentiation, communication, and exchange. In its first instance this process of recognition and new relations was only amongst the *Izbrisani* who found each other and, with some exceptions, created a community of care. Simultaneously, in a second instance, an insurgent citizenship was enacted, and support was garnered from outside of the embryonic *Izbrisani* movement. Todorivić sent letters to media, church, government and academic institutions. The Ljubljana Peace Institute started collecting and publishing *Izbrisani* stories on-line, in magazines and, latterly in books (Zorn and Lipovec Čebron 2008; Kogovšek et al. 2010). With the first Caravan of the Erased (see Figure 5.1), Amnesty International took notice. Their lawyers joined with Italian lawyers connected to the Ljubljana Peace Institute to help with the second Caravan of the Erased and the trip to the European Court of Human Rights in Strasbourg. Lawyers, activists, students and NGO workers made themselves available to erased people. Beginning in February 2006 an Erased Week was organized with events to highlight the plight and struggle of erased people. In 2013, Croatian theater director Oliver Frljic staged a play in Belgrade entitled *25,671* to raise awareness of the Slovenian erasure. In 2014, the play toured in Slovenia as part of that year's Erased Week (see Figure 6.1).

The 2014 Erased Week primarily highlighted events to raise awareness of the plight of children and young people, including a symposium and media event at Ljubljana University. These external connections were not just about staging and media events, however, they were crucial for developing community and establishing emotional citizenship amongst erased people.

Figure 6.1 Prior to a performance of *25,671* in Ljubljana, cast-members write the names of erased people. The step-ladders represent Triglav Mt. (see Figure 5.2)
Source: author

It is important to understand, as suggested by Agamben (1995) and Braidotti (2013), that *conatus* is embedded in *zoē*, and from bare life an emotional connection with others is possible. From this comes the potential for a political push in the world; and it is my contention that there is an aesthetics to this push that comprises who we are, what we say, how we say it, what we do, when, where and with whom, and there is also a pull that generates community from sharing, helping, and caring. For Braidotti (2013), as noted in previous chapters, *zoē*-centered egalitarianism is a platform for a post-humanist world that pushes back against the ideal of the Vitruvian man, and the excesses of rationality and reason. For me, *zoē* also comprises a political and transformative aesthetics that is grounded in a poetics of space, and some of its most dramatic and important political pushes come from young people (Aitken 2014). The idea of spatial poetics as a political push comes in part from Henri Lefebvre's (1991) notion that no community can gain political acumen without a 'trial by space'. For the *Izbrisani* youth, the previous chapter demonstrates a trial that was about mobility, movement, and being seen in public spaces that were heretofore dangerous. What is important to emphasize as these arguments move forward is that the young *Izbrisani* are connected in the world – to each other, to caregivers, to generations, to communities – even when, and particularly when, they are pushing back against all of it. It is impossible to overstate the evidence of connectivity from the stories in the last two chapters; the movements are not just about youth movements, because they involve adults, caregivers, teachers, institutions, and organizations in profound ways, and they are not about a single event or movement, as actions and activities take place over time and often over generations, nor is it just about focused activism because the movements also take place in the everyday and in routine activities. The idea of de-centering childhood and youth from a somewhat monadic bio-politics and re-establishing them as contextualized within a myriad of material and non-material, relations and dependencies ties in with Kraftl's (2014) idea of 'alter-childhoods', which explicitly attempts to imagine, talk about, and put into practice childhoods that differ from constructions through the last century and problematically since the UNCRC. Similarly to what I am trying to do here, albeit from a vastly different empirical perspective, Kraftl (2014: 219) elaborates the fluidity, hybridity and bio-power surrounding young people, with a specific focus on "intimacy, love, and the human-scale."

With the idea of youth relationalities and 'alter-childhoods' firmly established, and to the extent that the specific rise of *Izbrisani* youth as a political movement is a connected and contextualized trial by space and push back against neoliberal statehood and governance, this chapter raises the issue of stateless children and youth activism as political and global *causes célèbres*. I use this penultimate chapter, then, to broaden my discussion to four other youth social movements from elsewhere. Of course, my choice of raising certain movements and places while excluding others is somewhat arbitrary, but a certain degree of arbitrariness is inherent with comparison as an epistemological device for approaching social phenomena. My intent is to let these "concrete instances illuminate each other in their differences and affinities" (Erber 2016: 2). In the first instance, I look at perhaps the most specific case of young people 'outwith place'.[1] By taking over the streets in their school uniforms – by being, literally, out-of-their-place but still students (outwith) – the Chilean *Pingüinos* created a new space that caught the attention of the world. Second, I raise the issue of Roma families dispossessed of housing in Bucharest as an example of young people 'without space'. Third, in a different way, through *rolezinhos* (little strolls without shopping), in 2013 and 2014 black low-income young people claimed the spaces of upper middle-class shopping malls in Brazil. My fourth example briefly outlines the struggle of the DREAMers in the USA, their victories and obstructions to attain a place in the country within which they grew up. With these examples, and from my arguments in Chapter 3, I maintain that spatial rights at their simplest are about accommodating the presence of young people, and not to do so is a form of oppression. I am not attempting to bring these examples together to suggest some kind of universal understanding of erasure; the point is to let concrete instances from elsewhere illuminate each other in their differences and coalesce in their affinities. Throughout I nonetheless use the ideas of erasure raised in the last two chapters to elaborate the importance of interweaving young people as subjects and objects of rights, and how as societies we may push beyond rights to create a complex sustainable ethics through locatable feminist politics that recognize the multiplicities of a post-child moment. As I extend the Slovenian case with these very specific (and necessarily brief) youth protest examples from elsewhere in addition I push education (Chile), housing (Romania), consumerism (Brazil), and social security (USA). Out of the bare life and the spaces of exception occupied by young people it is possible to translate a message of hope.

Outwith place: Chile's *Pingüino* Revolution

In Spring 2006, Chilean elementary and high-school students took to the streets of Santiago to protest the continued privatization of education and an increasing disparity between rich and poor students. Within a few weeks the protests grew from a single march in Santiago to a nationwide campaign that placed half of the schools in Chile on strike or under occupation. The protests peaked on May 30 when 800,000 students took to the streets. Their rapid mobilization through text messaging took authorities by surprise and is generally regarded as the world's first social-media activated protest (Reel 2006). The protest became known as the *Pingüino* Revolution because the children wore their black and white uniforms while protesting, which looked from afar like a march of penguins. Sometimes the children took their desks from school and sat in them to fill the streets and block traffic. The sight of school uniforms and desks filling the streets suggest young people out-of-place (Aitken 2014: 108–117).

What is particularly interesting about the protest was the timing, in what Allison Bakamjian (2009: 2) calls a largely "demobilized" democratic era in

Chile. What she means by this is that there had been relative stability in Chile after Augusto Pinochet's dictatorship ended in 1992. The seeming new era of democracy inspired openness and communication that did not require activist mobilizations. Nevertheless, in 2006, the legacy of Pinochet's privatization of the education system – which epitomized the market-based nature of his right-wing economic and social policy reforms – remained a clarion call for many school pupils, particularly those from disadvantaged neighborhoods. There may have been more openness and communication in the new Chilean democracy, but young people were not feeling heard. Bakamjian (2009: 2) notes that "having attempted to change school privatization policies through more formal political means in the past, the students believed that a mass uprising was the only method that would gain sufficient government attention to make a difference."

To the degree that adults create a world in their own image, and to the degree to which that image is hugely flawed, Chile in the 1970s was a disturbing example of a nation that did not listen to its young people. According to Harvey (2005: 39) neoliberalism was birthed in 1973 with Pinochet's US backed military coup, which ousted the democratically elected Allende government. Salvador Allende's move towards socialism was up-ended, to be replaced by an economic model based on the work of Milton Friedman and his students at the University of Chicago. Since the 1950s, Chilean economics students were funded at the University of Chicago by the USA as part of a program to counter left-wing ideologies during the Cold War (Harvey 2005: 8). It is from here, and with these kinds of students, that Friedman's (1962) ideas of open markets and neoliberal governance took form. Once in power, Pinochet brought some of these young neoliberal economists – dubbed the 'Chicago Boys' – into his new government to prescribe a new ideological structure for his brave new world. Working with the International Monetary Fund (IMF), Chile's economy was restructured around theories of neoliberal capitalism. Friedman labelled this economic turnaround the 'Miracle of Chile'.[2] The Pinochet coup provided the USA with an opportunity to experiment with a neoliberal economic model based on Friedman's ideas of capitalism and freedom that embrace privatization, individuation and free-market global competition. The strategy of the Chicago Boys was threefold: to stabilize inflation, to initiate economic liberalization, and to privatize state-owned companies and enterprises (including, latterly, schools and hospitals). The policies initially were successful, at least in a limited but politically expedient way, and Chilean inflation and unemployment decreased through the late 1970s, but there was little net economic growth. Harvey (2005) emphasizes the short-lived nature of the Chilean economic turnaround – particularly with the debt crisis of 1982 – but the policies were in place and continued after Pinochet was gone. Ironically, after the debt crisis, the state controlled more of the economy than it had under Allende's socialist era and, as a consequence, neoliberal governance emerged from the economic policies. The seeming success of neoliberal economic policies in Chile created enthusiasm for market-based procedures that began to filter into social policy-making, including those that prescribed healthcare, welfare and education. In time, the processes of neoliberal governance moved relentlessly and consistently towards associated right-wing social doctrinarianism. I spend some time with the Chilean neoliberalization process because despite its demonstrated failure it became the model for economic and political reform in all the examples that follow.

Initiating a voucher system to pay for private, state-subsidized schools much like Charter School in the USA, the Chilean operation of schools was removed from the state and distributed to private enterprises and municipal governments. Fee-paying private schools were privately funded and operated, while municipal schools were free and operated by the municipal government. The private state-subsidized schools, on the other hand, were operated privately, but funded mostly by the government. As the funding was based on a voucher program, private schools received payment from individual families, while municipal and subsidized schools received a fixed amount per student (Bakamjian 2009: 14). One of the calls for action from the student protestors was against corruption or, at the very least, conflicts of interest, from the many government officials who owned stock in the newly privatized schools (Maria Jesús Sanhueza, Chilean Student Leader, 2012). This model extended a pernicious form of social doctrinarianism to the extent that schools were privatized and a free-market ethic was introduced into the entire education system, presaging a plan that began to find a foothold in North American and European education systems in the 1990s, which was expanded in the 2000s (Mitchell 2003, 2006, 2017).

The young people most at risk from the Chilean neoliberal governance, the young people with the most to lose, the people unable to vote or access the democratic process in any meaningful way because of their age, took to the streets. Lynn Staeheli and her colleagues (Staeheli et al. 2013, 2016) point out that the public and private status of actions is often equated with the spaces in which they occur such as homes, community centers, streets, planning departments or council chambers. Public policy makers and analysts tend to equate public actions with public spaces, and private actions with private spaces, but Marshall and Staeheli (2015) point to a much more complex mapping. Local youth actions become part of a community politics that loses power through a rigid and static conceptualization of scale, with homes and streets being the lowest point of entry and the state the highest. Staeheli tracks a locatable feminist politics that begins with care at the local level, at the scale of the community or even the kitchen, while simultaneously attacking public, national and international spheres of influence. This is where the *Pingüinos* began, while simultaneously branching onto streets and state government.

Although the leadership of the *Pingüinos* was exclusively adolescent, Chovanec and Benitez (2008: 41) note that there was a lot of "behind the scenes" guidance. Throughout the Pinochet dictatorship, when thousands of men went missing, radical feminist activism was constituted and practiced by women.

Chovanec and Benitez argue that while some of the young people involved
with the *Pingüinos* may have had no experience or little reference for revolutionary practices and political protest, there is a history of resistance from the
families of the young women who participated in the movement. Through
intergenerational learning, young girls "that played innocently 'under the
tables' as their mothers and grandmothers clandestinely planned and protested" gained insight into the worlds of repression and resistance without
having actually experienced it (Chovanec and Benitez 2008: 48). Thus, while
growing up in the 'demobilized' era of neoliberal democracy, the young
people of the *Pingüino* revolution not only felt free to express themselves
without the fear of authoritarian repression but also some had a distinctly
feminist familial background from which they knew how to organize. Chovanec
and Benitez (2008: 50) note that this ingrained social consciousness led to
greater social movement continuity in that the young women of the revolution
were influenced not only by their family members, but also through all of the
educators and activists in their lives.

Even as the *Pingüino* movement took off, government officials continued to
refuse to meet with students, and on May 21 2006 newly elected President
Michelle Bachelet stated that the movement was undemocratic (Bachelet
2006). A week later, the Minister of Education, Martin Zilic, agreed to meet
with the students but sent his sub-secretary instead. Feeling that they were not
being taken seriously, the students mobilized once more on May 30, just one
day after the failed meeting with Zilic. This marks the 800,000 strong protest
across the country, which continued for over a week as students negotiated
with a more attentive Zilic. A week later Bachelet went on state television
with a public announcement to the *Pingüinos* saying she realized that the
youth demands were justified and reasonable. After three months of struggle
and with many of their demands met and the creation of a Presidential
Advisory Council on Education with student representatives, the *Pingüino*
movement disbanded and normal classes were resumed with the proviso that
the government follow through with its promises.

The students' method of mobilization, which by-passed official institutional
channels, reveals a deep-seated belief in a system that no longer served their
interests. They are an exemplar of political pressure by youths too young to
vote but able to activate massive social connectivities to take back the streets.
Bakamjian (2009) focuses on the extraordinary nature of the *Pingüinos'*
mobilization amongst a group of young people (especially young women)
who had never experienced mass political demonstrations during the demobilized democratic era that followed the fall of Pinochet. The revolution is
remarkable to the extent that the young people had no model of street protest
and the use of social media propelled a movement against which the government
could not react in a timely fashion.

The *Pingüinos* maintained their protests from 2006 to 2014 as a result of
their increased capacity to use the education and political system to their
advantage even when it had initially failed them. This, I aver, is the epitome

of creating a space for *conatus* and *potentia*, and an emotional citizenship amongst people too young to vote. It may be argued further that the Bachelet administration took the form of a good-enough care-taker in creating a space for the activism to happen. Bachelet re-ran for and gained the Chilean presidency a second time in 2014, on a platform that involved, at least in part, education reform. She included several of the *Pingüinos* in her cabinet. Camila Vallejo Dowling was one of the prominent leaders of the protests in 2011. Her father and mother, Reinaldo Vallejo and Mariela Dowling, were both members of the Communist Party of Chile and were activists in the resistance during the Pinochet dictatorship. In 2006, Vallejo entered the University of Chile to study geography. There, she started forming ties with leftist students and became involved in politics, which led her to join the Chilean Communist Youth. Vallejo and Bachelet first met at a campaign event on June 15 2013. After Chile's national elections of 2013, Vallejo was elected to represent District 26 of La Florida with more than 43 percent of the votes, one of the highest victory margins of that election. Three other former student leaders were also sworn in as members of parliament during Bachelet's new administration.

The *conatus* of the *Pingüino* Revolution emanates from family ties, internal organization, counter-representation, social media, education, occupations, street marches and the bodies of the *Pingüinos*. The *Pingüino* movement was simultaneously internal and external, simultaneously physical space and virtual space, young people and their witnesses, and it was demonstrative of the equivalent and interconnected materialities and virtual potentialities of becoming through families, becoming by consensus, becoming online and becoming in the streets; they are not separate worlds or merely transitional forms but, rather, they are co-constitutive of material, non-material, private, public, common and individual interests. What is important here is that they began with disruption and surprise, garnered a witnessing public and went on to become the narrators of their own and their nation's history. It is important that young people surprise us, that they dislocate our complacency and get to rage against the status quo as a form of sustainability. It is particularly important at this time that they get to voice an opinion about the neoliberal educational reforms that promulgate profit and capitalism at the expense of their futures.

Without space: dispossession of Roma families in Romania

Following Romania's December Revolution in 1989, like in Slovenia and Chile, tensions arose between new democratic and neoliberal forms of governance, and older forms espousing socialism and social justice. In Romania, these tensions began with the new government taking responsibility for human rights violations attributed to the socialist era, including the nationalization of houses that deprived some Romanians of their property. These were mostly Romanians who fled the country during the communist era.

Tensions arose about how to make restitution without dispossessing those who now lived in the forfeited properties. In this part of the chapter, I focus on housing and social justice in Bucharest through a discussion of the displacement of Roma families following the restitution of nationalized properties, and the ways communities of care enabled some Roma families and young people to push back. As in Chile, education reform is also an issue in Romania (see Aitken and Arpagian 2018), but here I focus on housing security (see also Arpagian and Aitken 2018).

In addition to the complex laws underpinning restitution, and to the degree that Bucharest's historic neighborhoods are gentrifying, it is undeniable that neoliberal market processes are pushing aspects of low-income tenant dispossession. The erasure here, then, is housing dispossession for families who in actuality contrive strong connections to places, other marginalized people, and various institutions (church groups, non-governmental organizations (NGOs), charities and social organizations). The story, like the ones in Chile and Slovenia, is about the active and emotional citizenry that emanates from complex places and complex lives, and the ways that emotional citizenry creates communities of care, which sometimes enables push back against dispossession and exclusion from nation-statehood. The stories below of Roma families are derived from ethnographies and interviews conducted by Jasmine Arpagian in Fall 2015 and Summer 2016 (with some help from me). They are used to illustrate aspects of dispossession as it proceeds in precarious ways towards insurgent citizenship.[3]

Romania's progression of governments through the 1990s responded to internal and external pressures to redress human rights violations committed by the former communist regime (Olsen 2010). First, a privatization law passed in 1990 enabled three million tenants to purchase their apartments in state-constructed buildings at reduced prices; and in 1995 this possibility was extended to tenants of nationalized property (Stan 2006). Next, in 1991, legislation was passed to return vacant confiscated properties to former owners (Stan 2010). While the Parliament debated the fate of occupied properties not inhabited by their former owners, lower courts heard restitution claims and issued decisions. A law passed in 2001 protected occupants of returned buildings to some degree by requiring owners to extend rental contracts for five years, and rent could not be levied at more than a quarter of a household's annual income. Our interviews reveal that this requirement is inconsistently practiced and almost never enforced.

By 2012, 170,000 files were registered reclaiming properties across the country, but only nine percent had been resolved (Sultano 2012). According to the municipality's website, in October 2016 Bucharest still had more than 43,000 outstanding claims for property restitution. Following Judith Butler (2006), Arpagian and Aitken (2018) argue that the judicial and political landscape briefly described above forced property occupants into a state of liminal dispossession, where precarity was established and enhanced though destabilizing, dehumanizing processes.

There are clear similarities in Romania to the evolving neoliberal economies of other Eastern European countries that gained independence in the 1990s. Although the extent of the Slovenian erasure is not identifiable in Romania, there is nonetheless a context of dispossession exemplified by the creation of temporary, so-called *fără spațiu* (without space) identification cards, which require annual renewal. All Romanian citizens aged 14 years and over have identification cards. In 1997 a new card was issued that proved identity and established a home address or the house of a proprietor (e.g. a landlord) while no longer specifying a parental surname. The new card's removal of parental surnames de-emphasized lineage and focused instead on a specific address. This move highlights an aspect of housing insecurity for those who do not have a numbered street address, and particularly for families living on the street or in restituted apartment complexes. Although perhaps not as draconian as the dehumanizing six zeros *Izbrisani* found on their identification cards, the *fără spațiu* cards nonetheless identified their holders as decontextualized from a family (as orphans, without birth) and as dispossessed of a street address (as squatters, vagrants, without space). It seems clear that these provisional identification cards are part of a spectrum that reduces nation-statehood for the Roma we interviewed.

When we met her, Corina, a young mother of three, had lived informally for nearly ten years in restituted property:

> The temporary ID card is [now] something normal for us. Because there are so many cards, these are the people who don't have homes. Because to make a permanent identity card, you need residence documents.
>
> (July 2016 interview)

Corina clarifies that her ID card lists "an address without space. So, the street is written but it's not written that I stay at this or that number. So without space." Residential verification is required for permanent documents, which necessitates the applicant present a title of property, or a rental agreement or proof from the property owner indicating authorized occupancy. Owners of restituted properties rarely vouch for their occupants, who then receive provisional *fără spațiu* identification cards. Building owners often privately recognize their tenants but refuse to formalize the residence procedure because of concerns that any degree of official recognition (i.e. a rental contract or proof of occupancy) could legally complicate their property's future sale.

Mariana found herself *fără spațiu* when her salary could not cover her apartment's maintenance fee or the electricity: "My salary was not enough to support my children. I had four children, I was alone and young." She and her children moved into a run-down building occupied by other Roma. Her son Cornel points out: "We were raised this way, in poverty. There were days when we did not even have [anything] to put on the table. No, not days, weeks." He goes on to note that he had to miss a year of school because he was busy trying to help make ends meet. Mariana and her son were put in a space of domestic insecurity and legal liminality when "[t]he owner wrote a contract so we [can] stay for two years. But those two years passed and then she said you could stay here until the house collapses on you." Cornel interjects, with a hint of irony, "I take care of the house so it won't fall." When we finished our 2016 fieldwork, Mariana and her family had not been evicted; nonetheless they stay put with the realization that an eviction notice could be delivered any day.

This kind of dispossession sometimes initiates an active, emotional citizenry, enabling some Roma to rally against the injustices of emerging neoliberal forms of governance. With help from what Wright (2015) calls heterogeneous belongings that come about with 'feeling-in-common' through a complex assemblage of actors (including non-profit workers and charities), materials (wood, tarpaulin, and *fără spaţiu* cards) and places, families and individuals resisted or opposed their dispossession by staying in place (Butler and Athanasiou 2013; Curti et al. 2013; Gillespie 2016) and demanding alternative housing from the municipality. Emotional citizenry emanating from 'feeling-in-common' can lead to progressive politics (Wright 2015), but not always. Matthew Sparke (2013) notes that larger globalized processes can act repressively, which is why locatable feminist politics are never necessarily restricted to local communities.

As noted earlier, the locatable feminist politics that I am advocating emanate from new thinking about reproduction. Labor scholars and activists identify a new 'precariat' class in advanced capitalism that reflects the nature of uncertain and insecure work, which is temporary, part-time, and underemployed (Waitt 2011; Standing 2011). While this literature focuses on neoliberal landscapes of production, feminist geographers are also attuned to how precarity shows up in reproduction, and is lived in ways that are not bound by time, space or scale (Ettlinger 2007; Katz 2011; Clark 2015). Precarity is not just about loss of access to stable employment as suggested by Waitt (2011) and Standing (2011), then, it is about erosion of reproductive rights and a larger swathe of dispossession that de-territorializes "the genealogy of the proper(tied) subject." By suggesting this, Butler and Athanasiou (2013: 14) exacerbate the notion of a self-contained, liberal subject with particular status ascribed to propertied citizenship. The precarious neoliberal subject loses that status through state prescribed exceptions (Agamben 2001), mutations (Ong 2006), erasures, and dispossessions.

Although perhaps not as draconian as erased citizenship, improper (temporary, provisional) citizenship in Romania, it seems, is synonymous with provisional identification cards, and is connected to being 'without space'. Without space is not without complex place-based associations, including contacts in church groups and charities, and friendship networks of emotional citizenry. A refusal of impropriety (to be without property, space, place) then, is a "refusal to stay in one's proper place ... [and] signals an act of radical reterritorialization, which might certainly include remaining in specific places" (Butler and Athanasiou 2013: 21). Ana-Maria's story is in the first instance a move from security during the communist era, to dispossession and precarity in post-socialist Romania; second, it is a fight to stay close to (literally in a shack on the curb of) her demolished home; and third, with the help of friends in charities and NGOs, it is a provocative call on the municipality to house them as vulnerable citizens. After the court decided in favor of restitution to the former owner of the building she had lived in for 20 years, Ana-Maria and her neighbors became formal tenants 'with space' for five years under a new rental agreement. After those five years, Ana-Maria's new landlord sold the property to a developer attracted to its location in a neighborhood experiencing urban renewal. A few years after that, Ana-Maria's family and 26 others received eviction notices, and became 'without space'. Her home was razed to the ground. When asked how she felt watching the demolition of her home of 20 years, Ana-Maria responds:

> As you can imagine, it was our life's work. It was the childhood of my children who were raised and born there. And it's hard…Your labor goes and you don't have anywhere else to go, after a life of living in that house and you invested money. But it is not only about investment, because you make investments in any house you go. It is a matter of where the children grew up, there are memories there, things you lived over there. And it is hard, very hard. We do not have a lot of our belongings anymore. Now we have to begin from zero.
>
> (Ana-Maria, July 2016 interview)

When asked about the emotional impact of eviction, Ana-Maria's teenage son Andrei, was especially aware of the effects on his family's social network.

> Well, that is where we stayed since we were little. We were used to staying there. We were accustomed to everyone around us. We got along well with everyone. That is it. After they kicked us out, some went somewhere, others went somewhere else, and we were separated, among friends. And we're left with a few.
>
> (Andrei, August 2016 interview)

Andrei and Ana-Maria and her husband, and several neighbors, occupied their home's street in shacks for a year and a half in constant worry of removal without reasonable alternatives. Bodies on the street are precarious and demand regard (Butler and Athanasiou, 2013: i). Their vulnerability enabled Andrei and Ana-Marie to engage with friends to create "emotional geographies of belonging" (Askins 2016: 520–521), which helped build security and recognition in the face of confusing bureaucracies and unknown legalities. A sense of that which is possible (Butler 2006) empowers Ana-Maria to engage over time with friends in place-based encounters that promulgate inter-scale belonging and emotional citizenship through lines of connection and disconnection and fluid, hybrid identities and political allegiances.

In another corner of Bucharest's historic center, Corina's Roma neighbor recalls her visit to the city's housing administration in 2002, when her rent payment was declined because the property had been reclaimed by the former owner's heir. Although still recognized as an occupant, the reclamation stripped Corina of her right to formally and legally reside where she had lived for decades. The new owner did not draft a new rental contract, but did visit every few years, reassuring more than 20 families they would not be evicted, and advising them to collect firewood for winter.

> [We were here] since 2007. How I came here? We came here exactly like this, the owner reclaimed my parent's home. The owner came, we were evicted and we came here. And from here they are evicting us again... Here I found it through acquaintances, friends. I came and they let me stay out of pity, basically. I do not have documents here. I am staying on my own count. Because I did not have anywhere to go with three children after we were evicted.
>
> (Corina, July 2016 interview)

Over 20 (mostly Roma) families currently live here, uncertain and uncomfortable, but sheltered. About two dozen children play in the yard, only half of whom are enrolled in school. Their young parents hold precarious employment as unqualified laborers. Corina explains that she "had a work contract for four hours and [she] was working for eight," while her husband "works at a car wash. He doesn't have a salary, he works on tips and they schedule him non-stop; he stays three days, five days at work, day and night."

On a summer morning in 2016, a gendarme and local police show up and demand entry to the homes in the apartment, insisting that all families wait outside in the yard as rooms are searched for drugs. The property is reputed as a cheap sanctuary for neighborhood drug addicts; however most targeted families were uninvolved. Without a court order, the police acted on the pretext of abusive occupation by these families and cited that they were addressing "piling complaints from neighbors and reports of drug use and trafficking on the premises." Men and women were asked to stand separately while entry doors and storage shacks were knocked down. The families were given 48 hours to remove their belongings and themselves, but intervention from an experienced local volunteer activist enabled them to continue living in the apartment complex. A month after the eviction attempt, the families were still asking the local police for updates on their situation. They continued to wait with fear of becoming homeless in the winter. Despite this numbing uncertainty, Corina is tearfully grateful for the volunteers' unassuming and critical support:

> Thanks to that [volunteer], and I've said this so many times, because if he didn't come at that time, we would have been abusively removed, the way they wanted. When that boy came and said with what are you evicting

them? You came here to clean up, not to remove them. And no one said another word.

(July 2016, interview)

Emotional citizenry enables Corina and the other occupants in this instance to stay put, but with heightened feelings of precariousness.

Corina and Mariana stayed put. Ana-Maria and her family were forcefully removed but, as mentioned earlier, they did not move very far. After the eviction, many of Ana-Maria's neighbors were temporarily put up by family or friends. A few found apartments to rent. Having no housing alternative and refusing to accept this displacement, Ana-Maria and her family resisted by staying connected to a place directly adjacent to where the children were raised. Her decision was motivated by a perceived right to housing, and she felt that staying close elaborated that right.

We were not owners either, but we previously had documents [a rental agreement with the city's housing administration] ... That's why we stayed in place. **And because we stayed in place, we protested** ... Fine, it was also a matter of principle, but it was my right. To stay, for them to give me housing ... I paid the state for years.

(July 2016, interview, emphasis mine)

Support from local and international volunteers created for Ana-Maria the kind of emotional citizenry that Askins argues is inter-scaled and affective. Even with this support, the exposure to bureaucrats and the media was not easy for Ana-Maria and her family. When their shacks were ultimately bull-dozed in summer 2016, most of the resisting families received affordable housing in a new city-owned apartment reserved for evicted people on the periphery of Bucharest. When asked how he feels about their new home, Andrei responds: "The new house is a new life. We started a new life in it. We arranged it with what is needed. With what God was able to give us. And we are grateful we have someplace to rest our head."

When we last met with them, it was hard not to conclude that the outcome for Ana-Maria and her family was positive. They were happy with the newness of the social housing apartment, but they were nonetheless removed from the social networks and place-based attachments of their previous neighborhood. The emotional citizenry established there continues with church and charity connections in their old neighborhood. It is only a matter of time before Corina and Mariana, and their families, are moved away from the gentrifying core of Bucharest. The state responds with social housing, but it is also implicated in precarity and dispossession through the confounding housing restitution legislation and the de-humanizing *fără spațiu* cards.

It is fair to argue that the activism in Bucharest is not a primarily youth movement, although many young people are involved in the insurgency, the emotional citizenry and the creation of communities of care. I include it here to emphasize not only complex relationalities but also to provide a different register for childhood and youth that accommodates hope from everyday connectivities and dependencies, as well as the re-combinatorial powers of those connections, rather than simply from a blunt push against precarity and dispossession. As Kraftl (2008: 81) points out, in alignment with the larger thesis of this book, a focus on this kind of hope dispels simplistic and universalizing notions that link childhood to problematic notions of "hope, logic and futurity." The next example returns to a more singularly youth oriented movement, but in this instance I draw it into larger contexts of protest, history and racism.

Without consumption: Brazil's *rolezinhos* movement

The Romanian example exacerbates previous work on youth precarity that concentrates on productive activities such as access to employment (Waitt 2011; Standing 2011). When Chilean students pushed against privatization and issues of fairness in their education system, they also pushed against a neoliberal context of training for employment that, at least as it has evolved in UK, USA and Canadian education has come to focus mostly on standardizing testing to prepare young people for work in a neoliberal economy (Mitchell 2003, 2006, 2017). Nevertheless, and as Diane Ravitch (2013) has so powerfully argued, with privatization has come control by corporations more interested in short term profits from the schools in-and-of-themselves, rather than the long term preparedness and productivity of young people. It is not my intention to diminish the importance of productive activities, nor to give short-shrift to our understanding of the effects of unemployment, underemployment, and lack of job security on young people, but these concerns must be considered in conjunction with the ways neoliberal governance undermines reproduction. Consumption is the third part to Marx's understanding of economics, and it is to this that I turn in this third example of a youth movement.

In my arguments for a reconstituted, reproductive understanding of young people's geographies in the early 2000s, I raised the problematic image of "children as economic consumers and commodified packages" (Aitken 2001: 148). I noted that children's activities, at least on the Global North, are perhaps the most valuable source of market profit: not only through their parents but also increasingly through their own power in the market. This was not a new argument; Allison Diduck (1999: 128) recognized that to the degree children influenced household consumption it is not at all difficult to reconcile the identity "child" with that of "economic consumer." A related aspect of young people and the market is the way that they, and their activities, are commodified for market exploitation (Buckingham 2011).

My third example is about a moment in Brazil when relatively poor young people pushed back against exclusive spaces of consumption, although this characterization is to some degree mendacious, as the example is also a push for space, and perhaps also a drive against elitist and racist public policies. On

December 8, 2013 – at the beginning of the holiday shopping season – 3,000 young people entered and occupied São Paulo's *Metrô Itaquera* mall. Like the Chilean students, they organized through social media (in this case Facebook) and were part of the mall spaces before police or security could react. Moreover, and this is an important point, the mall was in a relatively exclusive shopping center in a rich part of São Paulo and the young people were mostly black and from *favelas*. Dubbed *rolezinhos* in the urban slang of São Paulo streets, the event was about poor young black people hanging out, and occupying an elite space of consumption for their own purposes. Those purposes were about dancing, flirting, playing music and, importantly, not shopping. Moreover, and to the surprise of the mall's security and management personnel, after the event was over there were almost no reports of shop-lifting or vandalism. Pedro Erber (2016) emphasizes that the *rolezeiros* and *rolezeiras* never claimed any particular political agenda as suggested by left-wing commentators, nor did they promote violence, theft, or vandalism as suggested by the conservative media. These events first came to the attention of the global press through Simon Romero's *New York Times* January 9, 2014 article that translated *rolezinhos* as 'little strolls', *The Guardian* called them flash-mobs (Watts 2014).

Over the next several months, *rolezinhos* popped up in other malls in São Paulo and Rio de Janeiro. On January 18, 2014, a *rolezinho* shut down the *JK Iguatemi* mall, another luxury shopping center in São Paulo. The reaction of the authorities, security teams and the malls' management was a predictable crack down. High-end malls got court injunctions to allow security personnel to bar the participants. Security guards wielding batons and pepper-spray chased the young people out of malls. Police showed up with tear gas and rubber bullets. In the Romero (2014) article 17-year-old high-school student Plinio Diniz asked "why don't they want us to go inside the malls? We have the right to have fun, but the police went too far." In the same article, Pablo Ortellado, a public policy professor from the University of São Paulo, notes that there are very few parks in the city and young dark-skinned people from the *favelas* "have been segregated from public space, and now they are challenging the unwritten rules." *Rolezinho* organizers argue that there are no politics underlying these activities and that they are simply flash-mobs and social gatherings. And yet, the lack of response to middle-class youth flash-mobs in *São Paulo* suggests that Ortellado's spatial politics hold sway in addition to the racism suggested by the authorities' response.

Erber (2016) ties the event of the *rolezinhos* to the larger unrest in Brazil during the previous year (the so-called Brazilian Spring) over increases in charges for public transportation in the run up to the 2014 FIFA World Cup, and argues that adjective 'little' suggests a characteristic ambiguity where no clear agenda is evident, there are no specific demands, and there is nonetheless deep-seated political unrest. This is in sharp contrast to the Brazilian Spring protests that were very specifically about a recent raise in public bus fares and a push for the long-term cause of free transportation for all. Erber

(2016: 3) points out that those most affected by the bus fare hike were poor people living in peripheral areas, "who often spent as much as five hours a day and 30 percent of their incomes in crowded buses and trains." For activist groups supporting the demonstrations against fare hikes, there was evocation of Brazil's 2001 *City Statute* that closely follows proclamations from the Right to the City movement, which has its origins in the work of French Marxist philosopher Henri Lefebvre.

Lefebvre's right to the city is not about universal rights or pseudo-rights, but rather it's "like a cry and a demand" (1996: 158). As such, he is concerned about the urban politics of the city's inhabitants, "formulated as a trans-formed and renewed right to urban life." Lefebvre offers a radical alternative that directly challenges and rethinks current urban structures of capitalism and liberal-democratic citizenship. His concepts highlight the importance of creating legal-political conditions to affirm a notion of social/affective citizenship that expresses new relations between urban inhabitants, politics and society. Brazil's City Statute focuses on citizen rights, municipalities' actions and the land market, and is a milestone in the implementation of the right to the city as a legal right. It aims to regulate urban policy so that the right to the city is recognized as a collective right through urban and spatial planning, environmental preservation, democratic management, social housing, and the regulation of informal settlements (Cordeiro et al. 2018). Erber (2016: 3–4) points out that during the 2013 unrest Sao Paulo activists supporting universal free bus fares in a public and collectively written document using language that closely corresponded to Lefebvrian sentiments, saying, "[t]hrough a process in which the population is always object rather than subject, transportation is ordered from above, according to the imperatives of the circulation of value. In this way, the population is excluded from the organization of its own everyday experience of the metropolis". The *rolezinhos* further exacerbated the claim that the City Statute was not working for everyone, and young black people in particular were not free to circulate in all urban spaces (Cordeiro et al. 2018), and particularly within what Erber (2016: 4) terms "Brazil's new fortresses of consumerism." The connection to the previous summer's unrest is important because the right to free circulation through public transportation that emerged at that time, evoking the City Statute, and "returned to the fore-front of the public debate in the guise of the purposeless circulation of strolling" (Erber 2016: 4). There is an important politics to strolling and it is within that politics that Erber tries to position the *rolezinhos*.

French poet Charles Baudelaire – dubbed by Walter Benjamin (1997) as the lyric poet of high capitalism – is famous for his construction of the strolling spectator on urban events, the *flâneur*. Baudelaire's *flâneur* ambled through the new fortresses of consumerism at the time, the Parisian shopping arcade. The creation of the arcade, essentially an enclosed walkway between two open streets, enabled the *flâneur* to stroll through enclosures and enter buildings publicly and gaze upon once private interiors. With 19th century arcades and 20th century malls, the open public street becomes an interior semi-public place (Aitken 1997). Of relevance to the *rolezinhos* is the work of feminists and post-colonial writers who elaborate and critique the gendered, classed and raced qualities of these urban spaces (cf. Buck-Morris 1986; Friedberg 1993). Baudelaire's *flâneur* was male and white, to which Friedberg (1993) rejoins that the female equivalent, the *flâneuse*, was a street-walker. Benjamin, in his unfinished *Arcades Project* (1982), writes that although the *flâneur* shares the situation and intoxication of the commodity, and although he is not necessarily militant, his privileges nonetheless require awareness and responsibility. Benjamin (1982: 448) admonished that "in our strolling, let us not forget our rights and our obligations as citizens. The times are necessitous; they demand our attention, all day long. Let us be *flâneurs*, but patriotic *flâneurs*." With that said, Erber (2016) cautions about making too much of the connection between the French *flâneurs* and the Brazilian *relezeiros*, because there is a different politics at work: for the latter the city is not open for contemplative spectatorship. What drives the passion for *rolezinhos* is not privileged spectatorship, but claims that revolve around the dispossession of city space.

At this point, it is worth contesting the wholesale invasion of middle- and upper-class malls by poor black youth. Leandro Beguoci (2014, cited in Erber 2016) notes that prior to becoming a global media event (and afterwards also), *rolezinhos* did not take place in malls in upper-class neighborhoods. *Rolezeiros*, Beguoci points out, mostly favor malls where they feel at home. This is a point that Erber uses to highlight an underlying problem of urban capitalist spatial development in its connection to a colonial context that pushes the poor and their malls to the urban periphery. In this sense, *rolezinhos* were laying claim to their own urban spaces, occupying the malls closest to them and, as such, reclaiming the right to the city stripped away from them by capitalist development.

As noted from Chapter 3, Winnicotian potential spaces are spaces of play, creativity and culture, and so it is worth pointing out that the *rolezinhos*, "with their tone of provocative playfulness (not unrelated to the proverbial playfulness of art), poked sensitive scars and open wounds of race and class relations in Brazil" (Erber 2016: 2). Like the Chilean *pingüino* demonstrations, which, with each proceeding year through 2014, got more and more carnivalesque as young people showed up in costume and their line-dancing took over Santiago's central square, the *rolezinhos* were lively, spirited and a juxtaposition to the fierce and sometimes violent reaction of security forces and police. And, like the 2011 Occupy Wall Street protests that began in New York City's Zuccotti Park and extended to many major cities globally, while the *roleizeros* and *roleizeras* made no specific demands, they touched a redounded nerve about the societal structures that created the problematic dissonance between the conspicuous consumption of the new Brazilian middle-classes and wholesale precarity of poor black youth. Erber's point is that the *rolezinhos*, when combined with the preceding Brazilian Spring and the protest that followed in 2015, suggest a push towards a more democratic society and the conservative backlash against working-class protests that brought to the fore some fundamental and unresolved tensions with the emergence of Brazil's new middle class.

What I want to underscore, pushing Erber's arguments a bit further and nonetheless conjoining them with *flânerie*, are the politics of presence suggested by *rolezinhos*. Not unlike the *pingüinos'* revolution, the *roleizeros* and *rolezeiras* are 'disturbing the sensible' through out-of-place, playful actions. As noted in Chapters 4 and 5, Rancière's (2009, 2010) work suggests the possibility of spontaneous popular uprisings through understanding the aesthetics of dissensus, of disturbing the seemingly sensible. Rancière view on aesthetics pushes traditional ideas of beauty, landscape and artistic sensibilities to a consideration of the "distribution of the sensible" (Rancière 2009: 1) in terms of relations "between what people do, what they see, what they hear, and what they know" (Rancière 2010: 15–17). Rancière argues that politics occurs when there is a disruption of a hegemonic or dominant mapping of the sensible. When young people occupy space, the aesthetics of the space change; there is a demand to be seen, to be visible. The *pingüinos* had very specific demands, whereas the *roleizeros* and *rolezeiras* by simply hanging were exhibiting a politics of presence.

The right to have rights: America's DREAMers

> [The DREAMers] were asserting their 'right to have rights': the right to have a public existence in a country that had banished them to the shadows.
>
> (Nicholls, 2013: 1)

Four undocumented students occupied the Arizona office of Senator John McCain on May 17, 2010 demanding residency status and the legal 'right to have rights' in the USA. They had come to the USA as children, had lived in the country most of their lives, and identified more as Americans than citizens of their birth countries. Without legal residency and with diminished rights, their situation was not much different from the *Izbrisani*. As with the *Izbrisani*, deportation was a possible outcome from their willingness to come out of the shadows and make a public statement. The McCain office event precipitated a flurry of activity throughout the USA, including more occupations of the offices of leading politicians, rallies, marches and video/text protests across various social media. Their agenda was to support the Development, Relief and Education for Alien Minors Act (DREAM Act). The Act, which provided a clear path to citizenship for qualifying alien minors in the USA, had been through various iterations since 2001 but all attempts had failed to get a majority vote. In what became known as the DREAMers movement, undocumented young men and women not only supported the senate bill, they also made themselves visible (and subject to deportation) and demanded to be recognized as viable human beings with rights to stay in the USA, and rights to a public and political life. Like the *Izbrisani* after Todorović's hunger strike at

Ljubljana Zoo, the McCain occupancy lit a touchstone for the DREAMer movement. The DREAMers realized that there were many other young people in the same situation, and like the *Izbrisani*, the embryonic political organization provided a forum for them to tell their stories.

In a comprehensive assessment of the DREAMers movement, Walter Nicholls (2013: 2) notes that the youth activism of May 2010 created a political movement that had heretofore not existed:

> There were no arguments, messages, or rhetoric to represent undocumented youths and their cause in the public sphere. There were no organizations to sustain their campaigns and interventions in public life. And there were few if any networks that allowed individual youths to connect to one another and create a sense of themselves as political beings.

In Spring 2010, DREAMer activists stepped forward to represent over one million people under the age of 25 years who had lived in the USA most of their lives, who had attended school and considered themselves Americans, but who remained undocumented. As epiphenomenal migrants, the DREAMers argued, they were not part of the decision-making to come to the USA illegally. As immigrants living most of their lives in the USA, they were engaged economically, civilly, and morally in the American system and they deserved the rights that accrued to citizenship. The DREAMers asserted that they were 'good' immigrants and deserving of permanent residency status, declaring themselves "undocumented, unafraid, and unapologetic," and not culpable for their illegal status. By so doing and at least while they were in the public eye, the DREAMers made themselves "un-deportable" (Nicholls 2013: 7).

The DREAMers movement continued after the first flurry of activities in May 2010 with the creation of community support groups, campus organizations, and a slew of social media appeals to organize and get involved. There was massive support from immigrant rights associations and other activist organizations such as the National Immigration Law Center (NILC) and the Center for Community Change (CCC). Over the next few years the DREAMers movement became the *cause célèbre* of the immigrant rights agenda. Nicholls (2013) and others (Schwiertz 2016) chart the connections between the DREAMers and the leading rights associations as they took a role in constructing the representation of the young people, connecting young activists, and training them to take their message into the public sphere. Similar to the beginning of the *Pingüinas* in Chile, who learned activism at the apron strings of their mothers and grandmothers, the DREAMers' activism was professionally amplified. Nicholls documents the rise of the movement and its dallying with professional activist groups, and how the young people eventually pushed back against what they thought were attempts to exercise control over the movement, to eventually assert autonomy and control over their own struggle (Schwiertz 2016). In so doing, there is an interesting course taken by the DREAMers from bare life without sacred and inalienable rights

(Agamben 1995: 75) and, specifically, excluded and in the shadows and labelled as doing something illegal, to a moral and political public high-ground with a public, powerful and legitimate voice that within a couple of years was virtually unassailable.

Nicholls' (2013) work provides a detailed account of the painstaking work required to produce a legitimacy for the DREAMers. Much like the *Izbrisani*, a meticulous process was required to break down the rhetorical and political opposition and "to find small cracks in the legal and moral systems of the country, making arguments for why their group deserved basic rights" (Nicholls 2013; 9–10). There was also unity and discipline amongst the DREAMers, which is uncharacteristic for long-term social movements. Successful movements like *Los Madres* in Argentina seeking justice for missing sons and daughters, for example, found the unity in their platform diluted because of internal power struggles (Bosco 2006). In Slovenia, the *Izbrisani* movement bifurcated three years after its formation when Todorović left the Association of Erased Residents (DIPS) to form the Civil Initiative of Erased Residents (CIIA, *Civilna Iniciativa Izbrisanih Aktivistov*) because of his disagreement with other early members. Internal divisions did not disrupt the DREAMers' main struggle as much as tensions with other immigrant activist groups, and, according to Nicholls, this was not necessarily an acrimonious or divisive tension. Rather, it strengthened the young activists' resolve. Helge Schwiertz (2016) argues further that the DREAMers moved beyond a narrowly defined struggle for legislative reform and citizenship status (which gained only limited success through deferred action) through the creation of undocumented communities that embraced intersectional power structures.

The original proposed DREAM Act of 2001 promised to place undocumented university students and young people performing community service on a path to citizenship. In the late 1990s, pressure to secure borders and increase deportation of illegal migrants resulted in unilateral enforcement measures that targeted egregious and non-egregious offenders alike, which put pressure on the Immigration and Naturalization Service (INS). Ironically, then, this pressure propelled INS officials to argue for deferred action – that is, temporary relief from deportation – for certain people on humanitarian and moral grounds. These calls for deferred action opened a window for the DREAM Act, which targeted a particular group of non-egregious offenders. The political and media work of the DREAMers was to replace the negative idea that they were non-egregious offenders with the positive idea that they were worthy of a path to citizenship. Although the original bill failed, strong support from key members of the House and Senate buoyed activism from the NILC and CCC to keep it alive over the next decade.

Nicholls (2013: 50–53) identifies three main themes that propelled a changing discourse of the DREAMers from 2001 onwards. First, was the embracing of American symbols, like fairness, hard work, self-determination and, of course, the American Dream. Rallies and marches fore-fronted American flags rather than Mexican flags. The DREAM Act was positioned as something that supported fairness and rewarded hard work, and the DREAMers were positioned as young people who were wholly American in their politics and cultural choices (hamburgers were preferred over burritos, and Bruce Springsteen over Andrés Contreras). Second, DREAMers positioned themselves not only as ordinary Americans but also as exceptional, the nation's 'best and brightest'. The image of the hard-working, tax-paying employee or, better, the 4.00 grade point average, college immigrant student, offset the stereotypical undocumented laborer who took American jobs and welched off the education and health system. Third, the stigma of unlawfulness was offset by the rhetoric of the epiphenomenal migrant who was not part of the decision-making to cross into the USA illegally. If DREAMers did not choose to cross the border illegally then they could not be held accountable for breaking the law.

By 2010, the DREAMers were frustrated by the actions of other immigrant action groups, and felt that larger immigrant interests for Comprehensive Immigration Reform were diluting their fundamental message. In April 2010 four undocumented students walked from Florida to Washington DC in an action dubbed the 'Trail of Dreams' that parodied the notorious Trail of Tears, and was reminiscent of the *Izbrisani* Caravan of the Erased to the European Court in Strasbourg. In Washington DC on May 1, a civil disobedience act was organized that resulted in the arrest of over a hundred supporters, including several members of Congress (Nicholls 2013: 80). This action fomented a substantial media reaction, which raised the DREAM Act for more debate. Following the success of the Trail of Dreams, supportive actions took place in many major US cities to push for the DREAM Act as a stand-alone bill. Many DREAMers felt that their push for a stand-alone bill was not taken seriously by activists in favor of comprehensive immigration reform, and so for them fairness came to also mean recognition as political equals (Nicholls 2013: 92). While maintaining support from, and complex ties with, national activist organizations, the DREAMers moved forward without ceding their autonomy. The first high profile action was the occupation of McCain's Arizona office, followed by the occupation of the Federal Building in Los Angeles, and congressional offices in Washington DC. There were many arrests, but the DREAMers had strategized to have some of their most solid legal cases incarcerated, thus offsetting as much as possible the risk of deportation.

The 2010 version of the DREAM Act failed because of a successful Republican filibuster in the Senate on December 18, 2010. This was a disappointing set-back, because the House had already passed a version of the bill and President Obama was generally supportive. By early 2011, any possibility for the DREAM Act or Comprehensive Immigration Reform was lost with the Republicans taking control of the House, and the Democrats losing their supermajority in the Senate. Nonetheless, as Nicholls (2013: 99) points out, the DREAMers had achieved the authority to express themselves in the public sphere, and it was a powerful voice that continued pressing the Obama administration through the next several years. In Spring 2012, President

Obama signed a memo calling for deferred action for undocumented immigrants who had come into the country as children. The Deferred Action for Childhood Arrivals (DACA) granted temporary legal status and work authorization for eligible immigrants while denying them access to healthcare. It is important to note that DACA did not provide a clear path to citizenship and could be revoked at any time by a sitting president. To take advantage of DACA, young people had to be younger than 31 years of age, had to have arrived in the USA before they were 17 years of age, and had to be a resident in the USA since 2007 with proof of education and no serious felonies or misdemeanors. At the time of writing, about 800,000 young people have signed up for DACA from the estimated 2.2 million of those who were eligible. Many did not sign up because of concerns that their personal information would be used against them if the DACA program was rescinded.

On 5 September, 2017, US Attorney General, Jeff Sessions, announced that the DACA program was being annulled, as Trump pledged he would do in his presidential election campaign. Trump stated that Congress could take up DACA if it wished, but he made it clear that this had to be in the context of broader immigration reform, including the building of a wall on the border and more limits on legal immigration. Dozens of protesters were arrested outside of Trump Tower in New York in one of many demonstrations nationwide, including Chicago and Los Angeles, in the aftermath of Sessions announcement.

Ala Sirriyeh (2017) has been studying the most recent work of the DREAMers against the onslaught of perhaps the most antagonist administration to their protracted cause. Highlighting the emotional work of the DREAMers, she makes the case for an affective citizenship that comprises the 'emotional transfer' so that the 'story of me' becomes the 'story of all of us' suffering together in solidarity (Sirriyeh 2017; see also Sirriyeh 2013). While the earlier rhetoric of the DREAMers focused on compassion – which foregrounded connections to the USA, and American symbolism – the new rhetoric focused on outrage and intersectionality.

Working primarily in California, Sirriyeh (2017) interviewed DREAMer organizers who told her that the optimism they had for working hard to attain the American Dream was now replaced with cynicism. They had worked hard, many attaining high educational levels, but now they were increasingly denied access to well-paying jobs or prestigious universities. For some, there was a realization that the American Dream was only for a particular kind of American whose storyline was racialized and classed. As one of Sirriyeh's (2017) respondents put it: "[legal] papers are not going to solve a lot of the oppression, a lot of the violence, exclusion and criminalization because as people of color we will never be accepted by this racist, homophobic, KKK country." This resonates with Schweirtz's (2016: 611) suggestion of a broadening and extension of the DREAMers as a social movement after 2011 to include larger contexts of insurgent citizenship against an oppressive homophobic and racist state. The problem that Sirriyeh raises with regard to the earlier categorization of DREAMers as hard-working, respectable, and deserving is their association with, and indexing to, whiteness. The communities of care to which DREAMers belonged became expansive and inclusive. Sirriyeh argues that with the move away from the idea of the DREAMer as wholly American and deserving of citizenship there is also a visceral connection with other disenfranchised groups such as LGBT as well as the larger undocumented community, and pushing a rhetoric of outrage against the unfairness of the US legal and immigration system. Further, she highlights the work of undocumented youth organizers in the traditional working-class neighborhood of Boyle Heights in Los Angeles, who also led the anti-gentrification campaign dubbed Defend Boyle Heights. Organizers argue that gentrification, criminalization and immigration enforcement are connected in ways that should propel insurgent citizenship from communities of care. The result of being thrown out of your home and your country are the same, irrespective of who is the perpetrator. With this recognition, affinity is drawn between Roma families displaced from their homes in Romania, and the erasure of young Slovenians, but also the rights of Chilean and Brazilian youth to take back their urban spaces.

In conclusion, some thoughts on reproduction and rights

That young marginalized people get to create and recreate themselves as they create and recreate their city is an important right to the degree that it emboldens a new model of governance and practice, which is progressive and potentially utopian rather than reactive and more-than-likely dystopian. To enact this kind of governance with urban poor and marginalized peoples is to push against the states of exception proffered by neoliberal governance. And to the degree that neoliberal economics is now roundly criticized as unworkable (Chakrabortty 2016), perhaps it is time for a brave new step that lets loose young, exceptional people and their creativities. The degree to which a neoliberal state-of-exception, whether by government fiat or economic duplicity, contextualizes young people and the spaces within which they live and from which they travel (or don't) is very much tied to how they see themselves, their capacities (in terms of *conatus/potentia* rather than individualized child agency) and how much there is push back against perceived injustices, then it is important for us to recognize their freedom to dislocate an old and tired political world. In the above examples, I intensify the complexities of *zoē*, and the connections between bare life and hopeful futures, which are tied to young people's mobility, exclusivity, entrapment, citizenship rights, affective citizenship and communities of care. The specific demands for rights and dignity in the spaces of exception that I began to describe in the previous chapter are highlighted as a problematical imaginary that lands on young people and their day-to-day contexts that then becomes amenable to locatable feminist politics, which are precisely about dislocation and the spaces through which surprise occurs. It may be argued that these feminist politics are simply a matter of scale, but they are not. The politics are firmly rooted in local places (cf. Gibson-Graham 2006; Grosz 2011; Braidotti, 2013) but the tree that arises from the roots may be 10 feet, 20 feet, or 50 feet tall, or its canopy may be shrouded in clouds, it is impossible to know. Perhaps, to continue the tree metaphor, the most important aspect of the roots are the rhizomes that move through earth strata in unpredictable ways, furnishing new and unexpected results. What is possible to know – what is clear from a feminist utopian hopefulness – is that the politics are, specifically, about reproduction writ large. I use the term reproduction here in a more expansive way, as the potential for young people to reproduce and remake themselves differently. The importance of the right to create and recreate themselves and their spaces is in the best interests of young people (and adults) and, as a consequence, the focus on spatial rights is not only about occupying spaces that are suitable for access to housing, livelihoods, consumption and education, but also about the right to stay put as well as the right of movement and mobility in safe and secure ways. This form of reproduction are locatable feminist politics that point to sustainable ethics. Sustainable ethics, in this sense and as I note throughout this book, are not about preservation of 'what is' into some unknown future, but rather they are about change and transformation in good and productive ways right now and only for this (perpetual) moment. As young people change their lives so too their contexts and environments are transformed if spaces are created to allow these transformations. This kind of imaginary works only in accordance with young people's power and position in society, the roles that are attributed to them, and the relations from which they derive freedom. For new youth imaginaries to work, the ideas of culture, geography and history must be more than a dystopian tool that lands a universal notion of who young people are in a particular locale, and it must be immune from the machinations of hierarchical state politics and the individualistic, market-driven strictures of neoliberal imagineering. The translation of a larger youth imaginary on particular locations is fraught with problems because the creation of imaginaries takes place not only between local, state and international institutions, but they also play out between the different spaces that young people occupy materially, non-materially, virtually and in their imaginations. In each of these spaces, and between all these institutions, different forms of translation occur.

Notes

1 As noted earlier 'outwith' is an old Scottish expression, but it has some extra resonance here. It suggests a positive tension between what is within and what is outside.
2 It is worth noting that the IMF in 2017 completely up-ended its views, not only noting the failure of neoliberal policies, but also arguing that economic programs that established tax programs to protect the wealth of the so-called one-percent were not sustainable (Chakrabortty 2016).
3 Some of this work is previously published in Aitken and Arpagian (2018) and Arpagian and Aitken (2018).

7 Sustaining young people through relational ethics

The idea of inalienable birth-rights was raised to sharp relief in 1739 when John Locke turned the pre-Enlightenment ideas of property ownership and entitlement on their head by advancing an idea that individuals gained property rights through their embodied labor. A lot of what I have been saying in the preceding pages revolves around the idea that the last two hundred years' focus on individual propertied (children acting properly?) and universal child rights (children properly cared for?) has obfuscated what precisely constitutes individualism, childhood, embodiment, labor, and doing. It is very clear that the universalizing of rights-based agendas is running into problems that go beyond what I outlined in Chapter 2. Apparently, corporations now have the rights of individuals, and individuals do not have the rights to their own embodied data. Today's technology and big-data, for example, turns what children's minds and bodies produce (birth-weights, heart-rates, test-scores, video-games played, social media friendships, and other propertied informatics) into properties that they do not own nor have rights over. Melinda Cooper (2008: 3) calls this extension "life as surplus," and questions

> [w]here does (re)production end and technical invention begin, when life is out to work at the microbiological or cellular level? What is at stake in the extension of property law to cover everything from the molecular elements of life (biological patents) to the biospheric accident (catastrophe bonds). What is the relationship between new theories of biological growth, complexity and evolution and recent neoliberal theories of accumulation?

When theories and practices like these become dogmatic they are translated into problematic rights agendas; for example universal child rights legitimize to some degree the right-to-life movement that pushes against women's rights to choose and rights over their bodies. This along with bodies reduced to their informational substrate, of course, is part of the excess of our posthuman moment, and it requires a new set of understandings, moralities and ethics.

A posthumanist perspective understands us as all-too-human, and as more than our corporeal selves, and it questions what precisely we can and should have rights over. Viewing children as relational doings, postchild advocates like Oswell (2013) and Murris (2016) argue, requires an understanding of their agencies and capacities in spaces of experience, experimentation and power: these spaces include the family, household, technology, social media, school, education, crime, criminality, health, medicine, play, consumer culture, political economies of labor, rights and political participation. These spaces do not recognize the divisions that seemingly encapsulated and cordoned past child-hoods. Rights and everyday politics in a post-global world are corporeal and technological, fluid, negotiable and relational, and they are tied to the ways that young people (and their relations with other people and things) create and recreate spaces of experience, experimentation and power.

To the degree that I am focused in this book on the more-than-child, I have less to say about the non-human and the un-human, although I do not shy away from discussion of inhuman practices. Young people are tied to things – nature, animals, technology, rooms, cafés, social media, games, banners and objects of protest – in important ways that are discussed only tangentially in the preceding chapters. Postchild researchers like Taylor and her colleagues (Taylor et al. 2012), Rautio and Winston (2015), and Murris (2016) deal much more specifically with the multiplicities of relations between children and things. I worry that some of these theorists lose an important political edge with a focus on the non-human that can reify a multiplicity of relationships to the degree that the important power relations are lost. Nonetheless, I am buoyed by those who continually circle-back to more structural post-colonial, racist and feminist politics. Murris (2016: 202), for example, articulates 'ethics of resistance' (from Lenz Taguchi 2010), which explicitly precipitate politically informed readings of the 'self' through picturebooks. More specifically, she evokes Deleuze and Guattari's (1994) notion of a line of flight for young children as a way to rupture and de-territorialize binary ways of knowing through the materialities of picturebooks. Perhaps even more pointedly, Affrica Taylor and her colleagues highlight contexts of indigenous politics through an analysis of very young children's relations to the non- and more-than-human world.

My project resides most comfortably in the more-than-human/more-than-child world rather than the non-human world. My lack of explicitly recognizing the non-human in this book does not detract from the implications of that presence in the biopolitics that frequently arise in the preceding chapters. My evocation of the postchild is one that attends more pointedly to human, family and institutional relations. I am intent upon a radically different conception of young people ousted from a child-centered world and contextualized through *zoē*, radical relationalities, insurgent and affective citizenship, non-heteronormative families, and ethics of care as well as an ethics of resistance. A child de-centered and unseated from their monadism is able to show up in multiple relationalities and political presence.

My focus on more-than-human postchild relationalities emanates from concerns over centuries of child-centered thinking, which converged with the

UNCRC on universal child rights and the acutely polemical 'best interests of the child'. When discussing the globalization and universalizing of children's rights agendas, Karen Wells (2015) notes that discourses move beyond protecting children from harm and acting in their best interests in problematic ways. In words that mirror Braidotti (2013), Wells (2015: 203) points out that in its global elaboration, children's rights are contrived from liberal ethics that hold inviable "the human as a subject who is universally a free, autonomous, rational, choosing individual." Wells goes on to note that the "normative model of contemporary childhood is, then, not simply about what it means to be a child, it is essentially about what it means to be human." Sentiments such as these propelled the post-UNCRC work of the 1990s and early 2000s that formed the foundation of the so-called new sociology of childhood (Jenks 1996; James et al. 1998). This new sociology drove the idea that children must be considered as 'separate beings' – with their own special needs, wants, experiences and rights – rather than 'becoming adults/becoming more than us'. This makes some sense because it proffers onto children and young people – as a group, as bearers or rights – a singular political acumen that is not derived from connections to adults. Of course, as with the course of women's rights through the 20th century, by lumping all children and young people together, intra-group distinctions, differences and intersectionalities are glossed over. In the middle chapters of this book I spend some time considering how universal rights based agendas came about for young people, noting that it is too simple to place blame on Enlightenment thinking, the UNCRC, and the liberal ethics of the Global North. As Wells (2015: 203) points out, the "universal subject that is at the heart of liberal theory does not and cannot exist because it presupposes that all humans share the same potential experience of the world." Children who create and re-create their own spaces do so in specific places and at particular moments.

Liberal ethics have never existed anywhere at any time, nor have they ever been an adequate utopian ideal, although Flax (1993) and other early feminist theorist on rights have argued that if these are all we have, then we must use them well and move forward as best we can on behalf of the best interests of women, children, and other minorities. Certainly it is laudable to use this perspective as a way of tackling social and spatial inequalities where they arise but I do not think that this is enough. In bringing together the examples of the last three chapters I have tried to exorcise global, universal ideals and liberal ethics while not losing their wilful use in specific places for singular purposes. With this book, however, I am trying to suggest another way forward, a way that dispels with vigor and assurance, the last vestiges of the unattainable Vitruvian ideal. What happens if we give up on children as monadic beings, with all the specific and singular rights that accrue to that position? My concern with the position of the UNCRC and the new sociology of childhood is that they do not untie connections to children 'becoming-the-same' as us, eventually. At some point, the monadic child becomes the monadic adult. One problem of the UNCRC's focus on the singularity of young people and the new sociology of children elaborating the importance of children 'being' rather than 'becoming' is seen in the plethora of contemporary childhood studies that include people in their twenties and thirties as young people. Coming from Deleuze and Guattari (1983, 1987), I prefer to think of young people as 'becoming-other', and giving up to them the space to become something different, something surprising, something unimaginable. Echoing Harvey's (2008: 23) sentiments on the rights to the city, and to restate the disarmingly simple supposition of this book, I prefer to give young people the right to create and recreate space and, by so doing, to recreate and recreate themselves and the world.

In the 1960s and 1970s, people living in Los Angeles bemoaned the rise of the power of the automobile and the turning of green space into a concrete and asphalt wilderness. Impersonal shopping malls with ample parking had taken over the consumptive spaces of downtown. School playgrounds were paved over and painted with lines to delimit specific activities and games. Flying into LAX airport was to traverse over mile after mile, block after block, of what seemed from the air to be ubiquitous grey urban slab. While adults lamented their loss of nature, community and place, a group of working-class pre-teens looked out from the tops of canyons on all the asphalt, tarmac and concrete with wonder and delight. They attached small wheels to narrow boards and created a new way to be in the city, a new mobility, a new way to exploit LA's corridors and parking lots, its drained swimming pools. For a short time, skateboarders became urban knights, the heroes of LA's byways and backstreets, grinding and jumping the sidewalks and benches. Their number grew and, before long, they were a noisy and boisterous danger to themselves and others. Laws were passed to exclude them from public places, skateboard parks were established to contain them, and the sport was legitimized on prime-time television; their clothes and lifestyles were commodified and sanitized. Nonetheless, skateboarding began with a dislocation, with a re-territorialization; an 11-year-old skateboarder looked out on the Los Angeles landscape with wonder and excitement, and claimed its marginal, moribund and abandoned spaces. That young person was not isolated or monadic, she was not on her own but connected to the non-material vastness of urban space that called out to her as an active part of her imagination and being and, in a moment, that space started to become something different and so did she. How, then, do we sustain a world that enables this kind of ethic to thrive? How do we create a world of potential and play for young people? How do we move away from rights agendas that are child-centered, turning rather to the idea of a young person "enmeshed in an immense web of material and discursive forces, always intra-acting with everything else" (Murris, 2016: xi)?

Throughout this book, with Braidotti (2006, 2013), Murris (2016), Oswell (2013) and others, I push the more-than-human postchild perspective as an alternative to liberal ethics, which leave children alone and impotent in the center of a world that is not of their making. Braidotti (2013) argues for a posthuman and post-anthropocentric ethics that focus on the missing people because, with Enlightenment a certain person was put forward as human (e.g. the Vitruvian man), and this person was not a child, or a woman or a skateboarder. Braidotti's neo-materialist philosophy of immanence posits all matter as one, as intelligent and as self-organizing. Braidotti's sustainable ethics come from Spinoza's monistic, relational understanding of God, the universe and us. God, according to Spinoza, is the natural world and everything in it, including us, in a multiplicity of interdependencies. Given this interdependency, Spinoza's ethics push against the notion of a Cartesian, mind/body split. The mind and the body are the same thing, Spinoza argues, they are just thought of in two different ways. Perhaps most importantly for the relational ethics I am trying to elaborate for young people, Spinoza argues that the mind/body cannot know its own thoughts/feelings better than it knows the ways in which its body is acted upon by other bodies and materialities. Further countering the mind/body split is the idea that we learn through and with our bodies, which Karen Barad (2012) characterizes as part of our intra-actions as a thing in relation to and influencing other things. Through intra-action, Spinozan thinkers like Barad and Braidotti argue that all things strive to persevere and continue. Spinoza calls this striving *conatus*, which as is suggested in the stories throughout this book, articulates the idea of living life to the full. It is the basis of sustainable ethics, and it is through postchild thinking that we get there.

The postchild is our historical and geographic condition, which is materially embedded and calls for the end of disciplinary purity. The postchild is multi-layered, nomadic, relational to human and non-human agents, and is mediated through technology. For this, argues Braidotti (2013), we need an adequate technology; a body/mind/thing map. This cartography is materially embedded, theoretically driven, and ethically progressive. Vital materialist neo-humanism suggests a way forward towards this cartography as ethics of sustainability that replaces the current liberal moral philosophy of children's rights. To get to that place, I find a suitable strategy in feminist politics of location, which scream from the empirical portions of the book. The distinct posthumanist character of the body/mind/thing map hinges on Spinoza's monist notion of difference, which posits difference through immanence rather than identity (Murris 2016: 110; see also Deleuze 2001). The idea of young people as beings propagated by the UNCRC and the new sociology of childhood assumes substance, monadism and political identity, which makes it a specific and passive object and a static and definable subject of rights. Alternatively, locatable feminist politics move from a 'freedom from' into action, doing and a 'freedom to', as Grosz (2011) points out, and they also, she goes on to offer, move us towards a radical rewriting of the singularities of modernity, which cannot be achieved by negating the past. Rather, the future and the past must come together in the perpetually unfolding present right here in this place, right now. This is the cornerstone of locatable feminist politics and the beginnings of sustainable ethics. In *The Coming Community*, Agamben (1993: III/7) describes a "whatever singularity" where "whatever" is not indifference but precisely a "being such that it always matters." By so doing, he moves beyond Lefebvre's (1996) notion of group rights through a trial-by-space, to describe an "inessential (anti-essential) commonality, a solidarity that in no way concerns an essence," a subject or an identity. By way of example, Agamben's beginning gambit is to describe love as something that is not "directed towards this or that property of the loved one (being blond, being small, being tender, being lame), but neither does it neglect the properties in favor of an insipid generality (universal love): The lover wants the loved one with all of its predicates, its being such as it is." This, I think, is precisely what Kraftl (2008) is after with his idea of childhood-hope (in the moment and from young people), which is radically different from the idea of some kind of universal hope emanating from the hopeless idea of children as our future. Like Braidotti's *conatus* and *potentia*, Agamben's coming community is emergent, it takes place; it is about love, intimacy and child-hope, and it has a locatable politics in communities of care.

This is precisely where I have tried to arrive, somewhere between specified universalism and locatable actions where no permission is given, and nothing is overcome. Rather, "truth is revealed only by giving space or giving place to non-truth – that is, as taking place of the false, as an exposure of its own innermost impropriety" (Agamben 1993: IV/13). Acceptance of paradoxes such as these – of the love and hate, the good and evil that reside within each of us as part of the without and the outwith, which Agamben (1993: IV/15) describes as an "innermost exteriority" – and the kind of vulnerability that bears with it an undeniable truth that foments the hope I describe as the on-going process of heart-work (Aitken 2009). Openness to this heart-work requires us, as adults, to know our intra-actions (with things, bodies, children) better so that we can set healthier boundaries but mostly so that we can let go lightly, and trust more fully that young people will do the right thing if they reside in a place that enables life to be fully lived. If I am still talking about rights, then it is about transformed and new rights to lifespace as a radical alternative that directly challenges and rethinks current structures of capitalism and liberal-democratic citizenship.

References

Abebe, Tatek and Anne Trine Kjørholt (2013). *Childhood and Local Knowledge in Ethiopia: Livelihoods, Rights and Intergenerational Relationships.* Trondheim, Norway: Akademika forlag.

Adams, Paul, Liela Berg, Nan Berger, Michael Duane, A.S. Neill and Robert Ollendorff (1971). *Children's Rights: Towards the Liberation of the Child.* New York: Praeger Publishers.

Agamben, Giorgio (1993). *The Coming Community.* Translated by Michael Hardt. London, Minneapolis: University of Minnesota Press.

Agamben, Giorgio (1995). *Homo Sacer: Sovereign Power and Bare Life.* Translated by Daniel Heller-Roazen. Stanford, CA: Stanford University Press.

Agamben, Giorgio (2001). *Means without End: Notes on Politics.* Translated by Vincenzo Binete and Cesare Casarino. London, Minneapolis: University of Minnesota Press.

Agamben, Giorgio (2005). *State of Exception.* Translated by Kevin Attell. Chicago and London: University of Chicago Press.

Aitken, Stuart C. (1997). Contesting the Mobilized Virtual Gaze. *Environment and Planning A: Society and Space,* 15: 113–126.

Aitken, Stuart C. (1998). *Family Fantasies and Community Space.* New Brunswick, New Jersey and London: Rutgers University Press.

Aitken, Stuart C. (2001). *Geographies of Children and Young People: The Morally Contested Spaces of Identity.* New York and London: Routledge.

Aitken, Stuart C. (2009). *The Awkward Spaces of Fathering.* Aldershot: Ashgate Press.

Aitken, Stuart C. (2014a). *The Ethnopoetics of Space and Transformation: Young People's Engagement, Activism and Aesthetics.* Aldershot: Ashgate Press.

Aitken, Stuart C. (2014b). Places of Interiority. In Paul Adams, Jim Craine and Jason Dittmer (editors) *Research Companion to Geographies of Media,* pp. 185–200. Aldershot: Ashgate Press.

Aitken, Stuart C. (2016). Locked-in-Place: Young People's Immobilities and the Slovenian Erasure. *Annals of the Association of American Geographers,* 106(2): 358–365.

Aitken, Stuart C. and Thomas Herman. (1997). Gender, Power and Crib Geography: From Transitional Spaces to Potential Places. *Gender, Place & Culture: A Journal of Feminist Geography,* 4(1): 63–88.

Aitken, Stuart C. and Don ColleyIII (2011). Spaces of Schoolyard Violence. In Michele Paludi (editor). *The Psychology of Teen Violence and Victimization,* pp. 83–105. Santa Barbara: Praeger Press ABC-CLIO.

Aitken, Stuart C. and Jasmine Arpagian (2018). Dystopian Spaces and Roma Imaginaries: The Case of Young Roma in Slovenia and Romania. In Andy Jonas, Byron Miller, Kevin Ward and David Wilson (editors). *Handbook on Spaces and Urban Politics.* New York and London: Routledge.

Aitken, Stuart C., Kate Swanson, Fernando Bosco and Tom Herman (2011). *Young People, Border Spaces and Revolutionary Imaginations.* New York and London: Routledge.

Aitken, Stuart C., Kate Swanson and Elizabeth Kennedy (2014). Independent Child Migrants: Navigating Relational Borderlands. In Spyros Spyrou and Miranda Christou (editors), *Studies in Childhood and Youth: Children and Borders,* pp. 214–242. Basingstoke: Palgrave MacMillan.

Archard, David (1993, 2nd edition 2004). *Children: Rights and Childhood.* New York and London: Routledge.

Arendt, Hannah (1962). *The Origins of Totalitarianism.* New York: Harcourt, Brace & Co.

Ariès, Philippe (1962). *Centuries of Childhood: A Social History of Family Life.* New York: Alfred A. Knopf.

Arpagian, Jasmine and Stuart C. Aitken (2018). Without Space: The Politics of Precarity and Dispossession in Post-Socialist Bucharest. *Annals of the American Association of Geographers.* Forthcoming, On-line.

Askins, Kye (2016). Emotional Citizenry: Everyday Geographies of Befriending, Belonging and Intercultural Encounter. *Transactions of the Institute of British Geographers,* 41(4): 515–527.

Bachelet, Michelle (2006). Address to the Nation, May 21, 2006, http://www.pre sidencia.cl/documentos/mensaje-presidencial-archivos/21Mayo2006.pdf   (accessed Feb. 17, 2013).

Bakamjian, Allison (2009). Chile's Penguin Revolution: Student Response to Incomplete Democratizationhttp://stonecenter.tulane.edu/uploads/Bakamjian_WEB-1312389920. pdf (accessed Feb. 13, 2012).

Balibar, Étienne (2012). The 'Impossible' Community of Citizens: Past and Present Problems. *Society and Space,* 30(3): 437–449.

Barad, Karen (2007). *Meeting the Universe Halfway: Quantum Physics and the Entanglement of Matter and Meaning.* Durham, NC: Duke University Press

Barad, Karen (2012). Intra-actions: An Interview with Karen Barad by Adam Kleinman. *Mousse,* 34: 76–81.

Barker, John, Peter Kraftl, John Horton and Faith Tucker (2013). The Road less Travelled– New Directions in Children's and Young People's Mobility. *Mobilities,* 4(1): 1–10.

Beguoci, Leandro (2014). Rolezinho e a desumanização dos pobres. htpp://www.oene. com.br/rolezinho-e-desumanizacao-dos-pobres/ (accessed Jan. 19, 2016).

Benjamin, Walter (1978). On the Mimetic Faculty. In P. Demetz and E. Jephcott (editors) *Reflections,* pp. 333–336. New York: Harcourt Brace.

Benjamin, Walter (1982). *The Arcades Project.* Translated by Howard Eiland and Kevin McLaughlin. Cambridge, MA and London: Harvard University Press.

Benjamin, Walter (1997). *Charles Boudelaire: A Lyric Poet in the Era of High Capitalism.* Translated by Harry Zohn. New York: Verso Press.

Bergson, Henri (1903/1999). *An Introduction to Metaphysics.* Translated by T.E. Hulme. Indianapolis/Cambridge: Hacket Publishing Company.

Berlant, Lauren (2011). *Cruel Optimism.* North Carolina: Duke University Press.

Berlant, Lauren (2012). *Desire/Love.* New York: Punctum Books

Berlant, Lauren (2015). Living in Ellipses. Keynote lecture given at *the International Emotional Geographies* conference, Edinburgh, Scotland.

Bethel School District vs Fraser No. 403 (1986). *Legal Information Institute*, Cornell University. https://www.law.cornell.edu/supremecourt/text/478/675 (accessed Dec. 28, 2017).

Blaise, Mindy (2005). *Playing it Straight: Uncovering Gender Discourses in the Early Childhood Classroom*. New York and London: Routledge.

Bosco, Fernando (2006). The Madres de Plaza de Mayo and Three Decades of Human Rights' Activism: Embeddedness, Emotions, and Social Movements. *Annals of the Association of American Geographers*, 96(2): 342–365.

Bosco, Fernando, Stuart C. Aitken and Tom Herman (2011). Women and Children in a Neighborhood Advocacy Group: Engaging Community and Refashioning Citizenship in a Border Town. *Gender, Place and Culture*, 18(2): 155–178.

Braidotti, Rosi (2006). The Ethics of Becoming Imperceptible. *Deleuze and Philosophy*, ed. Constantin Boundas, pp. 133–159. Edinburgh: Edinburgh University Press.

Braidotti, Rosi (2013). *The Posthuman*. Cambridge: Polity Press.

Buck-Morris, Susan (1986). The Flâneur, the Sandwichman, and the Whore: The Politics of Loitering. *New German Critique*, 39: 99–140.

Buckingham, David (2011). *The Material Child: Growing up in Consumer Culture*. London and New York: Polity Press.

Bunge, Bill and R. Bordessa (1975). *The Canadian Alternative: Survival, Expeditions and Urban Change*. Geographical Monographs, No. 2. Toronto: York University.

Burman, Erica (1994). *Deconstructing Developmental Psychology*. London and NY: Routledge.

Burman, Erica (1996). Local, Global or Globalized? Child Development and International Child Rights Legislation. *Childhood*, 3(1): 45–66.

Butler, Judith (2006). *Precarious Life: The Powers of Mourning and Violence*. London and New York: Verso Books.

Butler, Judith and Athena Athanasiou (2013). *Dispossession: The Performative in the Political*. London: John Wiley & Sons.

Brun, Cathrine, Piers Blakie and Mike Jones (2014). *Unravelling Marginalisation, Voicing Change: Alternative Geographies of Development*. Farnham: Ashgate Press.

Cantwell, N. (1992). The Origins, Significance and Development of the UNCRC. In S. Detrick (editor). *The UNCRC: A Guide to the Travaux Préparatories*, pp. 19–30. Dordrecht: Martinus Nijhoff.

Cigar, Norman (1995). *Genocide in Bosnia: The Policy of 'Ethnic Cleansing'*. College Station: Texas A & M Press.

Chakrabortty, Aditya (2016). You're Witnessing the Death of Neoliberalism – from Within. *The Guardian*, May. https://www.theguardian.com/commentisfree/2016/may/31/witnessing-death-neoliberalism-imf-economists (accessed Dec. 12, 2017).

Chovanec, DonnaM. and Alexandra Benitez (2008). The Penguin Revolution in Chile: Exploring Intergenerational Learning in Social Movements. *Journal of Contemporary Issues in Education*, 3(1): 39–57.

Clark, J. (2015). 'Just One Drop': Geopolitics and the Social Reproduction of Security in Southeast Turkey. In K. Meehan & K. Strauss (editors), *Precarious Worlds: Contested Geographies of Social Reproduction*. University of Georgia Press.

Cocoran, Steven (2015). Editor's Introduction. In Jacques Rancière, *Dissensus: On Politics and Aesthetics*, pp. 1–31. Bloomsbury: London and New Dehli.

Cooper, Melinda (2008). *Life as Surplus: Biotechnology and Capitalism in the Neoliberal Era*. Seattle WA: University of Washington Press.

Cordeiro, Adriana, Stuart C. Aitken and Sergio Mello (2018). Policy and Practice: Brazil's City Statute and Young People's Right to the City. In Afua Twum-Danso Imoh, Michael Bourdillon and Sylvia Meischner (editors) *Global Childhoods Beyond the North-South Divide*. Forthcoming.

Cunningham, Hugh (1995). *Children in Western Society since 1500*. London: Longman.

Curti, Giorgio Hadi, Jim Craine and Stuart C. Aitken (2013). *The Fight to Stay Put: Social Lessons through Media Imaginings of Urban Transformation and Change*. Stuttgart: Franz Steiner Verlag.

Curti, Giorgio Hadi, Stuart C. Aitken, Fernando J. Bosco and Denise Dixon Goerisch (2011). For Not Limiting Emotional and Affectual Geographies: A Collective Critique of Steve Pile's 'Emotions and Affect in Recent Human Geography'. *Transactions of the Institute of British Geographers*, 36(4): 590–591.

da Vinci, Leonardo (1883/repr.1970). 'Codex Leicester,' in *The Notebooks of Leonardo Da Vinci*, Vol. 1. New York: Dover Press, 179.

de Certeau, Michel (1984). *The Practice of Everyday Life*. Berkeley, CA: The University of California Press.

Deleuze, Gilles (1983). *Nietzsche and Philosophy*. New York: Columbia Press.

Deleuze, Gilles (1990). *Expressionism in Philosophy: Spinoza*. New York: Zone Books.

Deleuze, Gilles (2001). *Pure Immanence: Essays on A Life*. New York: Zone Books.

Deleuze, Gilles and Felix Guattari (1983). *Anti-Oedipus: Capitalism and Schizophrenia*. Minneapolis: University of Minnesota Press.

Deleuze, Gilles and Felix Guattari (1987). *A Thousand Plateaus: Capitalism and Schizophrenia*. Translated and with foreword by Brian Massumi. Minneapolis and London: University of Minnesota Press.

Deleuze, Gilles and Felix Guattari (1994). *What is Philosophy?*New York: Columbia University Press.

Diduck, Allison (1999). Justice and Childhood: Reflections on Refashioning Boundaries. In M. King (editor) *Moral Agendas for Children's Welfare*, pp. 125–137. London and New York: Routledge.

Dixon, Deborah (2009). Creating the Semi-living: On Politics, Aesthetics and the More-than-human. *Transactions of the Institute of British Geographers*, 34: 411–425.

Eisenstein, Zillah R. (1981). *The Radical Future of Liberal Feminism*. New York: Longman.

Elshtain, Jean B. (1990). The Family in Political Thought: Democratic Politics and the Question of Authority. In J. Sprey (editor) *Fashioning Family Theory*, pp. 51–66. Newbury Park, CA: Sage Publications.

Erber, Pedro (2016). The Politics of Strolling. *Latin American Perspectives*, 10: 1–16. doi:10.1177/0094582X16647717

Ettlinger, Nancy (2007). Precarity Unbound. *Alternatives*, 32: 319–340.

Fanon, Frantz (1967). *The Wretched of the Earth*. Harmondsworth: Penguin.

Fernando, Jude L. (2001). Children's Rights: Beyond the Impasse. *Annals of the American Academy of Political and Social Science*, pp. 8–25.

Filmer, Robert (1680). *Patriarcha, Or, the Natural Power of Kings*. Printed for Ric. Chiswell …, Matthew Gillyflower and William Henchman.

Flax, Jane. (1990). *Thinking Fragments: Psychoanalysis, Feminism, and Postmodernism in the Contemporary West*. Berkeley: University of California Press.

Flax, Jane (1993). *Disputed Subjects: Essays on Psychoanalysis, Politics, and Philosophy.* New York and London: Routledge.

Flax, Jane (2001). On Encountering Incommensurability: Martha Nussbaum's Aristotelian Practice. In James P. Sterba (editor), *Controversies in Feminism*, pp. 25–46. Lanham, Maryland: Rowman and Littlefield.

Foucault, Michel and Paul Rabinow (1997) *Ethics: Subjectivity and Truth: The Essential Works of Foucault, 1954–1984 Vol 1.* New York: New Press.

Fortas, Mr.Justice (1968) *Tinker v. Des Moines: Independent Community School District.* http://www.bc.edu/bc_org/avp/cas/comm/free_speech/tinker.html (accessed Nov. 20, 2011).

Foster, David (1994). Taming the Father: John Locke's Critique of Patriarchal Fatherhood. *The Review of Politics*, 56(4): 641–670.

Friedberg, Anne (1993). *Window Shopping: Cinema and the Postmodern.* Berkeley: The University of California Press.

Friedman, Milton (1962). *Capitalism and Freedom.* Chicago: Chicago University Press.

Front, Sonia and Katarzyna Nowak (2010, editors), *Interiors: Interiority/Exteriority in Literary and Cultural Discourse.* Newcastle-Upon-Tyne: Cambridge Scholars Publishing.

Frønes, Ivar (1994). Dimensions of Childhood. In Jens Qvortrup, Margarita Bardy, Giovanna Sgritta and Helmut Winterberger (editors) *Childhood Matters: Social Theory, Practice and Politics*, pp. 145–164. Aldershot, UK: Avebury Press.

Gaïni, Firouz (2013). *Atlantic Wayfarers: Essays on Young Faroe Islanders.* Steiten, Norway: Lap Lambert Academic Publishing

Gagen, Elizabeth (2000a). An Example to Us All: Child Development and Identity Construction in Early 20th Century Playgrounds. *Environment and Planning A*, 32(4): 599–616.

Gagen, Elizabeth (2000b). Playing the Part: Performing Gender in America's Playgrounds. In Sarah Holloway and Gill Valentine (editors), *Children's Geographies: Playing, Living and Learning*, pp. 213–229. London and New York: Routledge.

Gagen, Elizabeth (2008). Reflections of Primitivism: Development, Progress and Civilization in Imperial America, 1898–1914. In Stuart C. Aitken, Ragnhild Lund and Anne Trine Kjørholt (editors). *Global Childhoods: Globalization, Development and Young People*, pp. 16–28. London and New York: Routledge.

Geertz, Clifford (1973). Deep Play: Notions on a Balinese Cockfight. In *The Interpretation of Culture*. New York: Basic Books.

Gibson-Graham, J.K. (2006). *A Postcapitalist Politics.* Minneapolis: The University of Minnesota Press.

Giddens, Anthony (1998). *The Third Way.* London and New York: Polity Press.

Gillespie, T. (2016). Accumulation by Urban Dispossession: Struggles over Urban Space in Accra, Ghana. *Transactions of the Institute of British Geographers*, 41(1): 66–77.

Gleeson, B. and N. Sipe (2006). Reinstating Kids in the City. In *Creating Child Friendly Cities*, pp. 1–10. New York: Routledge.

Gopnik, Alison (2016). *The Gardener and the Carpenter: What the New Science of Child Development Tells Us About the Relationship Between Parents and Children.* New York: Farrar, Straus and Giroux.

Graham, Steve and Stephen Marvin (2001). *Splintering Urbanism.* London and New York: Routledge.

Grosz, Elizabeth (2011). *Becoming Undone: Darwinian Reflections on Life, Politics and Art.* Durham, NC: Duke University Press.

Gregorčić, Marta (2008). Phantom Irresponsibility, or Facism in Disguise. In Jelka Zorn and Uršula Lipovec Čebron (editors) *Once Upon an Erasure: From Citizens to Illegal Residents in the Republic of Slovenia*, pp. 115–132. Ljubljana: Študentska založba.

Guðmundsdóttir, M. L., J. Sigfússon, Á. L. Kristjánsson, H. Pálsdóttir and I. D. Sigfúsdóttir (2010). *The Nordic Youth Research among 16 to 19 year old in Åland Islands, Denmark, Faroe Islands, Finland, Greenland, Iceland, Norway and Sweden.* Icelandic Centre for Social Research and Analysis.

Hall, G. Stanley (1904). *Adolescence; its Psychology and its Relations to Physiology, Anthropology, Sociology, Sex, Crime, Religion and Education.* New York: D. Appleton and Company.

Hall, G. (1909). *Fifty Years of Darwinism: Modern Aspects of Evolution.* Centennial addresses in honor of Charles Darwin, before the American Association for the Advancement of Science, Baltimore, Friday, January 1, 1909. New York: H. Holt and Company

Hayfield, Erika (2017). Exploring Transnational Realities in the Lives of Faroese Youngsters. *The Journal of Nordic Migration Research*, 7(1): 3–11.

Harari, Yuval Noah (2015). *Sapiens: A Brief History of Humankind.* New York: Harper Collins.

Haraway, Donna (1988). Situated Knowledges: The Science Question in Feminism and the Privilege of Partial Perspectives. *Feminist Studies*, 14(3): 575–599.

Hartman, Sven (2009). *Janusz Korczak's Legacy: An Inestimable Source of Inspiration: Lectures on Today's Challenges for Children.* Strasbourg, France: Office of the Commissioner for Human Rights, Council of Europe. https://www. coe.int/t/commissioner/source/prems/PublicationKorczak_en.pdf (accessed Dec. 14, 2017).

Harvey, David (2000). *Spaces of Hope.* Berkeley and Los Angeles: University of California Press.

Harvey, David (2005). *A Brief History of Neoliberalism.* Oxford and New York: Oxford University Press

Harvey, David (2008). The Right to the City. *New Left Review*, 53: 23–40.

Hazelwood School District vs Kuhlmeier 484 US 260 (1988) No. 86–836. Justia, US Supreme Court. https://supreme.justia.com/cases/federal/us/484/260/case.html (accessed Dec. 28, 2017).

Henricks, Thomas (2015). *Play and the Human Condition.* Urbana, Chicago, Springfield: University of Illinois Press.

Herman, Arthur (2001). *How the Scots Invented the Modern World.* New York: Three Rivers Press.

Holston, John (2009). *Insurgent Citizenship: Disjunctions of Democracy and Modernity in Brazil.* Princeton, NJ: Princeton University Press.

Homes, Amy, Susan Solomon, Annie Wild, Chris Creegan and Paul Bradshaw (2014). *Views & Experiences of the Children's Hearings System Research with Children, Young People & Adults.* Scottish Centre for Social Research: Edinburgh. http:// www.chscotland.gov.uk/media/68051/Views-and-Experiences-of-the-Childrens-Hea rings-System-v1-0.pdf (accessed June 6, 2017).

Hopkins, Peter and Rachel Pain (2007). Geographies of Age: Thinking Relationally. *Area*, 29(3): 287–293.

Horton, John and Peter Kraftl (2006). Not just Growing Up, but Going On: Materials, Spacings, Bodies, Situations. *Children's Geographies*, 4(3): 259–276.

Howard, Neil (2014). On Bolivia's New Child Labour Law. *Beyond Trafficking and Slavery.* https://www.opendemocracy.net/beyondslavery/neil-howard/on-bolivia%E2%2 80%99s-new-child-labour-law (accessed June 20, 2015).

Huizinga, Johan. (1955). *Homo Ludens: A Study of Play Elements in Culture.* Boston: Beacon.

Human Rights Watch (2009). *US: Jumpstart Ratification of Women's Rights Treaty.* On 30th Anniversary, Progress Flagging on Obama's Pledge of Support. https://www.hrw.org/news/2009/12/18/us-jumpstart-ratification-womens-rights-treaty (accessed Dec. 28, 2017).

Hume, David (1739/1955). *A Treatise on Human Nature: Being an Attempt to Introduce the Experimental Method of Reasoning in Moral Subjects.* Edited with an analytical text by L. A. Selby-Bigge. London: Oxford at the Clarendon Press.

In re Gault, 387 USA 1 (1967). Supreme Court of the United States Decision. Appeal from the Supreme Court of Arizona no. 116 argued: December 6, 1966 – decided: May 15, 1967.

Isin, Engin F. (2008). Theorizing Acts of Citizenship. In E.F. Isin and G.M. Nielsen (editors) *Acts of Citizenship.* London and New York: Zed Books.

Isin, Engin F. (2014). *Citizens Without Frontiers.* London and New York: Bloomsbury.

James, Alison, Chris Jenks and Alan Prout. (1998). *Theorizing Childhood.* New York: Teachers' College Press.

Jalušič, Vlasta and Jasminka Dedič (2008). The Erasure – Mass Human Rights Violation and Denial of Responsibility: The Case of Independent Slovenia. *Human Rights Review,* 9: 93–108.

Jenks, Chris (1982). Constructing the Child. In C. Jenks (editor), *The Sociology of Childhood: Essential Readings.* London: Batsford.

Jenks, Chris (1996). *Childhood.* London: Routledge.

Johannesen, Aksel V. (2017). Prime Minister of the Faroese Islands on the Occasion of 'The population of the Faroe Islands reaches 50,000'. http://www.government.fo/news/news/the-population-of-the-faroe-islands-reaches-50-000/ (accessed Sept 21, 2017).

Johannesen, Aksel V., Høgni Hoydal and Poul Michelsen (2017). *Solidarity, Self-sufficiency and Liberty*: Coalition Agreement between the Social Democratic Party (Javnaðarflokkurin), Republican Party (Tjóðveldi) and Progressive Party (Framsókn). http://www.government.fo/the-government/coalition-agreement/ (accessed Sept 21, 2017).

Kallio, Kirsi and Jouni Häkli (2011a). Are there Politics in Childhood? *Space and Polity,* 15(1): 1–34.

Kallio, Kirsi and Jouni Häkli (2011b). Tracing Children's Politics. *Political Geography,* 30: 99–109.

Kallio, Kirsi and Jouni Häkli (2013). Children and Young People's Politics in Everyday Life. *Space and Polity,* 17(1): 1–15.

Katz, Cindi (2004). *Growing Up Global.* Minnesota: Guilford Press.

Katz, Cindi (2011). Accumulation, Excess, Childhood: Towards a Countertopography of Risk and Waste. *Documents d'Analisi Geografica,* 57(1): 67–72.

Katz, Cindi and Janice Monk (1993). *Full Circles: Geographies of Gender over the Life Course.* New York and London: Routledge.

Kingsbury, Paul and Steve Pile (2014). *Psychoanalytic Geographies.* New York and London: Routledge.

Kirby, Kathleen M. (1996). *Indifferent Boundaries.* New York: The Guilford Press.

Kjørholt, Anne Trine (2003). Creating a Place to Belong. Girls' and Boys' Hutbuilding as a Site for Understanding Discourses on Childhood and Generational Relations in a Norwegian Community. *Children's Geographies*, 1(2).

Kjørholt, Anne Trine (2008). Childhood as a Symbolic Space: Searching for Authentic Voices in the Era of Globalization. In Stuart C. Aitken, Ragnhild Lund and Anne Trine Kjørholt (editors), *Global Childhoods: Globalization, Development and Young People*, pp. 29–42. London and New York: Routledge.

Kogovšek, Neža (2010). The Erasure as a Violation of Legally Protected Human Rights. In Neža Kogovšek, Jelka Zorn, Sara Pistotnik, Uršula Lipovec Čebron, Veronika Bajt, Brankica Petkovič and Lana Zdravkovic (editors). *The Scars of the Erasure: A Contribution to the Critical Understanding of the Erasure of People from the Register of Permanent Residents of the Republic of Slovenia*, pp. 83–140. Peace Institute: Metelkova 60, 1000 Ljubljana, Slovenia.

Kogovšek, Neža, Jelka Zorn, Sara Pistotnik, Uršula Lipovec Čebron, Veronika Bajt, Brankica Petkovič and Lana Zdravkovic (2010). *The Scars of the Erasure: A Contribution to the Critical Understanding of the Erasure of People from the Register of Permanent Residents of the Republic of Slovenia*. Peace Institute: Metelkova 60, 1000 Ljubljana, Slovenia.

Kolmerten, Carol (1990). *Women in Utopia: The Ideology of Gender in the American Owenite Communities*. Bloomington, IN: Indiana University Press.

Korczak, Janusz (1929/2009). *The Child's Right to Respect: Janusz Korczak's Legacy Lectures on Today's Challenges for Children*. Strasbourg, France: Office of the Commissioner for Human Rights, Council of Europe. https://www.coe.int/t/comm issioner/source/prems/PublicationKorczak_en.pdf (accessed Dec. 14, 2017).

Kraftl, Peter (2006). Building an Idea: The Material Construction of an Ideal Childhood. *Transactions of the Institute of British Geographers*, 31(4): 488–504.

Kraftl, Peter (2008). Young People, Hope, and Childhood-Hope. *Space and Culture*, 11(2): 81–92.

Kraftl, Peter (2013a). Beyond 'Voice', Beyond 'Agency', Beyond 'Politics'? Hybrid Childhoods and some Critical Reflections on Children's Emotional Geographies. *Emotion, Space and Society*, 9: 13–23.

Kraftl, Peter (2013b). *Geographies of Alternative Education: Diverse Learning Spaces for Children and Young People*. Bristol, UK: Policy Press.

Kraftl, Peter (2014). Alter-Childhoods: Biopolitics and Childhoods in Alternative Education Spaces. *Annals of the American Association of Geographers*, 105(1): 219–237.

Krishnan, M. (2014). Advancing Backwards: Bolivia's Child Labor Law. *Council on Hemispheric Affairs*. http://www.coha.org/advancing-backwards-bolivias-child-labor-la w/ (accessed June 20, 2015).

Kuhelj, Alenka (2011). Rise of Xenophobic Nationalism in Europe: A Case of Slovenia. *Communist and Post-Communist Studies*, 44: 271–282.

Lacan, Jacques (1978). *The Four Fundamental Concepts of Psychoanalysis*. Translated by A. Sheridan. New York: W.W. Norton.

Laclau, Ernesto (1990). *New Reflections on the Revolution of Our Time*. London: Verso.

Lefebvre, Henri (1996 [1968]). The Right to the City. In *Writings on Cities*. Oxford: Blackwell.

Lefebvre, Henri (1991). *The Production of Space*. translated by Donald Nicholson-Smith. Oxford: Blackwell.

Lenz Taguchi, H. (2010). *Going Beyond the Theory/Practice Divide in Early Childhood Education*. New York and London: Routledge.

Lester, Stuart and Wendy Russell (2010). Children's Right to Play: An Examination of the Importance of Play in the Lives of Children Worldwide. Working Paper No. 57. The Hague, The Netherlands: Bernard van Leer Foundation.

Lillis, Mike, Rafael Bernal and Rebecca Savransky (2017). Trump Rescinding DACA Program. *The Hill*. http://thehill.com/latino/348848-sessions-says-DACA-to-end-in-six-months (accessed Dec. 1, 2017).

Locke, John (1693). *Some Thoughts Concerning Education*. A. & J. Churchill at the Black Swan in Paternoster-row.

Locke, John (1690/1824). *Two Treatises on Government*. London: C. & J. Riverton et al., and Edinburgh: Stirling and Slade.

Lopatka, Adam (2007). Introduction from the Chairman/Rapporteur of the Working Group on a draft convention on the rights of the child. *Legislative History of the Convention on the Rights of the Child*, Vol. 1, pp. xxxvii–xliv. New York and Geneva: Office of the United Nations High Commissioner for Human Rights. http://www.ohchr.org/Documents/Publications/LegislativeHistorycrc1en.pdf (accessed Dec. 7, 2017).

Louv, Richard (2005). *Last Child in the Woods: Saving Children from Nature-Deficit Disorder*. San Francisco: Algonquin Books.

Mackenzie, Suzanne (1989). Restructuring the Relations of Work and Life: Women as Environmental Actors, Feminism as Geographic Analysis. In Audrey Kobayashi and Suzanne Mackenzie (editors). *Remaking Human Geography*, pp. 40–61. Boston: Unwin Hyman.

Maldonado-Torres, Nelson (2016). Outline of Ten Theses on Coloniality and Decoloniality. frantzfanonfoundation-fondationfrantzfanon.com/article2360.html (accessed Dec. 6, 2017).

Marshall, David (Sandy) (2013). 'All the Beautiful Things': Trauma, Aesthetics and the Politics of Palestinian Childhood. *Space and Polity*, 17(1): 53–73.

Marshall, David and Lynn A. Staeheli (2015). Mapping Civil Society with Social Network Analysis: Methodological Possibilities and Limitations. *Geoforum*, 61: 56–66.

Marston, Sally, J.P. Jones and Keith Woodward (2005). Human Geographies Without Scale. *Transactions of the Institute of British Geographers*, 30(4): 416–432.

Massey, Doreen (2005). *For Space*. New York and London: Routledge.

Mekina, Igor (2007). Izbris Izbrisa (The Erasure of the Erasure). *Časopis za kritiko znanosti*, 35(228): 157–170.

Mekina, Igor (2014). Personal Interview conducted in March 2014.

Mitchell, Katharyne (2003). Educating the National Citizen in Neoliberal Times: From the Multicultural Self to the Strategic Cosmopolitan. *Transactions of the Institute of British Geographers*, 28(4): 387–403.

Mitchell, Katharyne (2006). Neoliberal Governmentality in the European Union: Education, Training and the Technologies of Citizenship. *Society and Space*, 24: 389–407.

Mitchell, Katharyne (2017). Changing the Subject: Education and the Constitution of Youth in the Neoliberal Era. In Tracey Skelton and Stuart C. Aitken (editors). *Theories and Concepts: Establishing Geographies of Children and Young People* (Geographies of Children and Young People Vol. 1 of 12). Springer Major Reference Work: Springer Publishing Company.

Mooney, G. (1998). *Urban 'Disorders'. Unruly Cities?: Order/Disorder*. New York and London: Routledge.

Moosa-Mitha, Mehmoona (2005). A Difference-Centred Alternative to Theorization of Children's Citizenship Rights. *Citizenship Studies*, 9(4): 369–388.

Moosa-Mitha, Mehmoona (2017). The Political Geography of the 'Best Interests of the Child'. In Tracey Skelton and Stuart C. Aitken (editors), *Theories and Concepts: Establishing Geographies of Children and Young People* (Geographies of Children and Young People Vol. 1 of 12). Springer Major Reference Work: Springer Publishing Company.

Morse v. Frederick, 551 U.S. 393 (2007) No. 06–278 Justia: US Supreme Court. https://supreme.justia.com/cases/federal/us/551/393/ (accessed Dec. 28, 2017).

Murris, Karin (2016). *The Posthuman Child: Educational Transformation through Philosophy and Picturebooks*. New York and London: Routledge.

Nast, Heidi (2000). Mapping the 'Unconscious': Racism and the Oedipal Family. *Annals of the Association of American Geographers*, 90(2): 215–255.

Nicholls, Walter J. (2013). *The DREAMers: How the Undocumented Youth Movement Transformed the Immigrant Rights Debate*. Stanford, CA: Stanford University Press.

NSPCC: National Society for the Protection of Cruelty to Children (2018). What We Do: Every Child is Worth Fighting For. UK national charity incorporated by Royal Charter. https://www.nspcc.org.uk/what-we-do/ (accessed Jan. 4, 2018).

Nussbaum, Martha (1997). *Cultivating Humanity: A Classical Defense of Reform in Liberal Education*. Cambridge, MA: Harvard University Press.

Nussbaum, Martha (2001). In Defense of Universal Values. In James P. Sterba (editor), *Controversies in Feminism*, pp. 3–24. Lanham, Maryland: Rowman and Littlefield.

Olsen, T. D. (2010). *Transitional Justice in Balance: Comparing Processes, Weighing Efficacy*. US Institute of Peace Press.

Ong, Aihwa. (2006). *Neoliberalism as Exception: Mutations in Citizenship and Sovereignty*. Durham NC: Duke University Press.

Oswell, David (2013). *The Agency of Children: From Family to Global Human Rights*. Cambridge: University of Cambridge Press.

Owen, Robert (1816/1972). *A New View of Society*. London: Macmillan Press Ltd.

Owens, Patricia (2009). Reclaiming 'Bare Life'?: Against Agamben on Refugees. *International Relations*, 23: 567–582.

Paine, Thomas (1791/1970). *The Rights of Man and Other Writings*. London: Heron Books.

Paul, S., N. Lee, Y. Clement, K. So and Louis Leung (2015). Social Media and Umbrella Movement: Insurgent Public Sphere in Formation. *Chinese Journal of Communication*, 8: 356–375.

Parekj, S. (2014). Beyond the Ethics of Admission: Stateless People, Refugee Camps and Moral Obligations. *Philosophy and Social Criticism*, 40: 645–663.

Petersen, Tóra (2016). *Mental Health among Youth in The Faroe Islands: Who is Responsible? What is Being Done?* Nordic Centre of Welfare and Social Issues.

Pimlott-Wilson, Helena and Sarah Marie Hall (2017). Everyday Experiences of Economic Change: Repositioning Geographies of Children, Youth and Families. *Area*, 49(3): 258–265.

Pollock, Linda (1983). *Forgotten Children: Parent–Child Relations, 1500–1900*. Cambridge: Cambridge University Press.

Postman, Neil (1982). *The Disappearance of Childhood*. New York: Delacourt Press.

Proudfoot, Jesse (2015). Anxiety and Phantasy in the Field: The Position of the Unconscious in Ethnographic Research. *Environment and Planning D: Society and Space*, 33(6): 1135–1152.

Prigge, Walter (2008). Reading the Urban Revolution: Space and Representation. In Kanishka Gooneward, Stefan Kipfer, Richard Milgrom and Christian Schmid (editors) *Space, Difference and Everyday Life*, pp. 46–61. New York and London: Routledge.

Rancière, Jacques (2004). *The Flesh of Words. The Politics of Writing*. Translated by C. Mandell. Pali Alto, CA: Stanford University Press.

Rancière, Jacques (2005). From Politics to Aesthetics. *Paragraph*, 28(1): 13–25.

Rancière, Jacques (2008). *Hatred of Democracy*. Translated by S. Corcoran. London: Verso.

Rancière, Jacques (2009). The Aesthetic Dimension: Aesthetics, Politics, Knowledge. *Critical Inquiry*, Autumn: 1–19.

Rancière, Jacques (2010). The Aesthetic Heterotopia. *Philosophy Today*, 54: 15–25.

Rancière, Jacques (2015). *Dissensus: On Politics and Aesthetics*. Bloomsbury: London and New Delhi.

Rautio, Pauliina and Johan Winston (2015). Things and Children in Play: Improvisation with Language and Matter. *Discourse: Studies in the Cultural Politics of Education*, 36(1): 15–26.

Ravitch, Diane (2013). *Reign of Error: The Hoax of the Privatization Movement and the Danger to America's Public School*. New York: Alfred A. Knopf.

Rawls, John. (1971). *A Theory of Justice*. Cambridge, MA: Harvard University Press.

Ricoeur, Paul (1981). *Hermeneutics and the Human Sciences: Essays on Language, Action and Interpretation*. Edited and translated by John B. Thompson. Cambridge: Cambridge University Press.

Reel, Monte (2006). Chile's Student Activists: A Course in Democracy. *The Washington Post*, Saturday, November 5.

Reynaert, Didier, Ellen Desmet, Sara Lambrechts and Walter Vandenhole (2015). Introduction: A Critical Approach to Children's Rights. In Walter Vandenhole, Ellen Desmet, Didier Reynaert and Sara Lambrechts (editors), *Routledge International Handbook of Children's Rights Studies*, pp. 1–23. New York and London: Routledge.

Riis, Jacob (1890). *How the Other Half Lives*. New York: Charles Scribner and Sons.

Romero, Simon (2014). Brazil's Latest Clash with its Urban Youth Takes Place in the Mall. *New York Times*, January 9. https://www.nytimes.com/2014/01/20/world/am ericas/brazils-latest-clash-with-its-urban-youth-takes-place-at-the-mall.html (accessed Dec. 12, 2017).

Rousseau, Jean-Jacques (1762). *Émile*. In *The Émile of Jean-Jacques Rousseau*. Translated and edited by William Boyd (1962). New York: Columbia University Bureau of Publications.

Šalamon, Neža Kogovšek (2016). *Erased: Citizenship, Residence Rights and the Constitution in Slovenia*. Frankfurt am Main: PL Academic Research.

Sanders, Mike (2001). *Women and Radicalism in the Nineteenth Century: Specific Controversies*. New York and London: Routledge.

Schechner, R. (1993). *The Future of Ritual: Writing on Culture and Performance*. London and New York: Routledge

Schwiertz, Helge (2016). Transformations of the Undocumented Youth Movement and Radical Egalitarian Citizenship. *Citizenship Studies*, 20(5): 610–628.

Sirriyeh, Ala (2013). *Inhabiting Borders, Routes Home*. Farnham, England: Ashgate.

Sirriyeh, Ala (2017). 'Felons are also Family': Undocumented Youth, Solidarity and Citizenship. Paper presented at the IVth International Conference on Geographies of Children, Youth and Families, Loughborough, England.

Skelton, Tracey (2008). Children, Young People, UNICEF and Participation. In Stuart C. Aitken, Ragnhild Lund and Anne Trine Kjörholt (editors) *Global Childhoods: Globalization, Development and Young People*, pp. 165–181. New York and London: Routledge.

Sparke, Matthew (2013). From Global Dispossession to Local Repossession: Towards a Worldly Cultural Geography of Occupy Activism. In Nuala C. Johnson, Richard Schein and Jamie Winders (editors). *The Companion to Cultural Geography*, pp. 387–408. London and New York: John Wiley and Sons.

Staeheli, Lynn and David Hammett (2013). 'For the Future of the Nation': Citizenship, Education, and Nation in South Africa. *Political Geography*, 32: 32–41.

Staeheli, Lynn and Alex Jeffrey (2017). Learning Citizenship: Civility, Civil Society, and the Possibilities of Citizenship. In Kirsi Kallio, Sarah Mills and Tracey Skelton (editors) *Geography of Children and Young People* (Vol 7): *Politics, Citizenship, and Rights*, pp. 497–514. New York: Springer.

Staeheli, Lynn and Caroline Nagel (2018). Narrating the Palimpsestic Space of Post-Conflict Cities. *Environment and Planning A*. Forthcoming.

Staeheli, Lynn, Kafui Attoh and Don Mitchell (2013). Contested Engagements: Youth and the Politics of Citizenship. *Space and Polity*, 17(1): 88–105.

Staeheli, Lynn, David J. Marshall and Naomi Maynard (2016). Circulations and the Entanglements of Citizenship Formation. *Annals of the Association of American Geographers*, 106: 377–384.

Staunaes, Dorthe (2016). Notes on Inventive Methodologies and Affirmative Critiques of an Affective Edu-future. *Research in Education*, 96(1): 62–70.

Stan, L. (2006). The Roof over our Heads: Property Restitution in Romania. *Journal of Communist Studies and Transition Politics*, 22(2): 180–205.

Stan, L. (2010). Romania: In the Shadow of the Past. *Central and Southeastern European Politics since 1989*: 379–400.

Standing, Guy. (2011). *The Precariat: The New Dangerous Class*. London: Bloomsbury Academic.

Stearns, Peter (2017). History of Children's Rights. In Martin D. Ruck, Michele Peterson-Badali and Michael Freeman (editors) *Handbook of Children's Rights: Global and Multidisciplinary Perspectives*, pp. New York and London: Routledge.

Stone, Lawrence (1974). The Massacre of the Innocents. *New York Review of Books*, Xxi(18), November 27.

Sultano, M. (2012). Despagubirea fostilor proprietari. Scandal monstru intre Guvern, ANRP si Asociatiile de proprietary abuziv deposedati. *Gandul*. April 24. http://www.gandul.info/stiri/despagubirea-fostilor-proprietari-scandal-monstru-intre-guvern-anrp-si-asociatiile-de-proprietari-abuziv-deposedati-9559437 (accessed June 20, 2015).

Sutton-Smith, B. (1997). *The Ambiguity of Play*. Harvard, MA: Harvard University Press.

Swerdlow, Amy, Renate Bridenthal, Joan Kelly and Phyllis Vine (1989). *Families in Flux*. New York: Feminist Press.

Taylor, Affrica and Veronica Pacinini-Ketchabaw and Mindy Blaise (2012). Children's Relations to the More-Than-Human World. *Contemporary Issues in Early Childhood*, 13(2): 81–85.

Thompson, N. A. (2014). Bolivian Children as Young as 10 Years Old Being Put to Work to 'Solve' National Poverty. *Latin Post*, July 17, 2014. http://www.latinpost.com/articles/17394/20140717/bolivian-children-young-10-years-old-being-put-work-solve.htm (accessed June 20, 2015).

Tinker vs De Moines School District (1968). Landmark Supreme Court Ruling On Behalf Of Student Expression. https://www.aclu.org/other/tinker-v-des-moines-la ndmark-supreme-court-ruling-behalf-student-expression (accessed Dec. 28, 2017).

United Nations (1989). *Convention on the Rights of the Child (UNCRC)*. www.unesco. org/education/pdf/CHILD_E.PD (accessed Nov. 29, 2013).

United Nations (2009). *Children's Rights at a Cross-Roads*. A Global Conference for The 20th anniversary of the UN Convention on the Rights of the Child. http://www. childwatch.uio.no/events/conferences/2009-cwi-conference-addis-ababa.html (accessed Nov. 29, 2013).

Waitt, L. (2011). A Critical Geography of Precarity. *Geography Compass*, 3(1): 412–433.

Watts, Jonathan (2014). Brazilian Flashmob Forces Upmarket Shopping Mall to Close. *The Guardian*. January 19. https://www.theguardian.com/world/2014/jan/20/ brazilian-flashmob-shopping-mall-closes-rolezinho (accessed Dec. 12, 2017).

Wells, Karen (2015). *Childhood in a Global Perspective*. 2nd edn. Cambridge: Polity Press.

Wilson, Adrian. (1980). The Infancy of the History of Childhood: An Appraisal of Philippe Ariès. *History and Theory*, 19(2): 132–154.

Winnicott, Donald W. (1953). Transitional Objects and Transitional Phenomena. *International Journal of Psychoanalysis*, 34: 89–97.

Winnicott, Donald W. (1965). *The Family and Individual Development*. New York: Basic Books.

Winnicott, Donald W. (1971). *Playing and Reality*. London: Tavistock.

Winnicott, Donald W. (1975). *Through Pediatrics to Psycho-analysis*. New York: Basic Books.

Winnicott, Donald W. (1988). *The Child, The Family and the Outside World*. London: Penguin.

Wood, Denis and Robert Beck (1994). *The Home Rules*. Baltimore and London: The Johns Hopkins Press.

Woodyer, Tara (2012). Ludic Geographies: Not Merely Child's Play. *Geography Compass*, 6(6): 313–332.

Wright, S. (2015). More-than-human, Emergent Belongings: A Weak Theory Approach. *Progress in Human Geography*, 39: 391–411.

Yezer, Caroline (2011). How Not to be a Machu Qari (Old Man): Human Rights, Machismo, and Military Nostalgia in Peru's Andes. In Dorothy Hodgson (editor), *Gender and Culture at the Limit of Rights*, pp. 120–134. University of Pennsylvania Press.

Young, Iris Marion (1990). *Justice and the Politics of Difference*. Princeton NJ: Princeton University Press.

Young, Iris Marion (2006). Education in the Context of Structural Injustice: A Symposium Response. In Mitja Sardoc (editor) *Citizenship, Inclusion and Democracy: A Symposium on Iris Marion Young*, pp. 91–101. Malden and Oxford: Blackwell.

Yuval-Davies, Nora. (1999). Ethnicity, Gender Relations and Multiculturalism. In R.J. Torres, L.M. Miron and J.X. Inda (editors), *Race, Identity & Citizenship: A Reader*, pp. 112–125. Oxford: Blackwell.

Zdravković, Lana (2010). The Struggle Against the Denial of Citizenship as a Paradigm of Emancipatory Politics. In Neža Kogovšek, Jelka Zorn, Sara Pistotnik, Uršula Lipovec Čebron, Veronika Bajt, Brankica Petkovič and Lana Zdravkovic (editors), *The Scars of the Erasure: A Contribution to the Critical Understanding of the Erasure of People from the Register of Permanent Residents of the Republic of Slovenia*, pp. 257–277. Peace Institute: Metelkova 60, 1000 Ljubljana, Slovenia.

Žižek, Slavoj (2006). *The Parallax View.* Cambridge, MA: MIT Press.

Žižek, Slavoj (2010). *Living in the End Times.* London and New York: Verso.

Žižek, Slavoj (2012). *Interrogating the Real.* London and New York: Continuum.

Žižek, Slavoj (2014). *Event: Philosophy in Transit.* Kindle E-book. London and New York: Penguin.

Zorn, Jelka (2010) Registered as Workers, Erased as non-Slovenes: The Transition Period from the Perspective of Erased People. In Neža Kogovšek, Jelka Zorn, Sara Pistotnik, Uršula Lipovec Čebron, Veronika Bajt, Brankica Petkovič and Lana Zdravkovic (editors). *The Scars of the Erasure: A Contribution to the Critical Understanding of the Erasure of People from the Register of Permanent Residents of the Republic of Slovenia*, pp. 19–46. Peace Institute: Metelkova 60, 1000 Ljubljana, Slovenia.

Zorn, Jelka (2011). From Erased and Excluded to Active Participants in Slovenia. In Brad K. Blitz and Maureen Lynch (editors), *Statelessness and Citizenship: A Comparative Study on the Benefits of Nationality*, pp. 66–84. Cheltenham, UK and Northampton USA: Edward Elgar.

Zorn, Jelka and Uršula Lipovec Čebron (2008). *Once Upon an Erasure: From Citizens to Illegal Residents in the Republic of Slovenia.* Ljubljana: Študentska založba.

Zucker, Eve (2013). Figures of Modernity: the Village Police Chief. In Joshua Barker, Erik Harms and Johan Lindquist (editors), *Figures of Southeast Asian Modernity*, pp. 75–90. University of Hawai'i Press.

Zupančič, Boštjan (2016). Preface. In Neža Kogovšek Šalamon. *Erased: Citizenship, Residence Rights and the Constitution in Slovenia.* Frankfurt am Main: PL Academic Research.

Index

Milton Keynes UK
Ingram Content Group UK Ltd.
UKHW040053071024
449327UK00019B/520